SpringerBriefs in Mathematics

SpringerBriefs present concise summaries of cutting-edge research and practical applications across a wide spectrum of fields. Featuring compact volumes of 50 to 125 pages, the series covers a range of content from professional to academic. Briefs are characterized by fast, global electronic dissemination, standard publishing contracts, standardized manuscript preparation and formatting guidelines, and expedited production schedules.

Typical topics might include:

A timely report of state-of-the art techniques A bridge between new research results, as published in journal articles, and a contextual literature review A snapshot of a hot or emerging topic An in-depth case study A presentation of core concepts that students must understand in order to make independent contributions.

SpringerBriefs in Mathematics showcases expositions in all areas of mathematics and applied mathematics. Manuscripts presenting new results or a single new result in a classical field, new field, or an emerging topic, applications, or bridges between new results and already published works, are encouraged. The series is intended for mathematicians and applied mathematicians. All works are peer-reviewed to meet the highest standards of scientific literature.

Titles from this series are indexed by Scopus, Web of Science, Mathematical Reviews, and zbMATH.

Kaïs Ammari • Fathi Hassine • Luc Robbiano

Stabilization for Some Fractional-Evolution Systems

Springer

Kaïs Ammari
Department of Mathematics
University of Monastir
Monastir, Tunisia

Fathi Hassine
University of Monastir
Monastir, Tunisia

Luc Robbiano
Laboratoire de Mathématiques
Université de Versailles Saint-Quentin
Versailles, France

ISSN 2191-8198   ISSN 2191-8201   (electronic)
SpringerBriefs in Mathematics
ISBN 978-3-031-17342-4   ISBN 978-3-031-17343-1   (eBook)
https://doi.org/10.1007/978-3-031-17343-1

This Springer imprint is published by the registered company Springer Nature Switzerland AG
The registered company address is: Gewerbestrasse 11, 6330 Cham, Switzerland

Contents

Chapter 1
Introduction

In recent years, fractional calculus has been increasingly applied in different fields of science [40, 55, 60]. Physical phenomena related to electromagnetism, propagation of energy in dissipative systems, thermal stresses, models of porous electrodes, relaxation vibrations, viscoelasticity, and thermoelasticity are successfully described by fractional differential equations [32, 39, 42, 43]. Fractional calculus allows for the investigation of the nonlocal response of mechanical systems, and this is the main advantage when compared to the classical calculus.

Fractional derivatives provide an excellent instrument for the description of memory and hereditary properties of various materials and processes. This is the main advantage of fractional derivatives in comparison with classical integer-order models, in which such effects are in fact neglected. The advantages of fractional derivatives become apparent in modeling mechanical and electrical properties of real materials, as well as in the description of rheological properties of rocks, and in many other fields.

Fractional integrals and derivatives also appear in the theory of control of dynamical systems, when the controlled system or/and the controller is described by a fractional differential equation [3, 28, 52]. Integer-order derivatives and integrals have clear physical interpretation and are used for describing different concepts in classical physics. For example, the position of a moving object can be represented as a function of time, the object velocity is then the first derivative of the function, and the acceleration is the second derivative. Fractional derivatives and integrals, being generalization of the classical derivative and integrals, are expected to have an even broader meaning. Unfortunately, there is no such result in the literature until now.

Fractional calculus includes various extensions of the usual definition of derivative from integer to real order [31], including the Riemann–Liouville derivative, the Caputo derivative, the Riesz derivative, the Weyl derivative, etc. In this book, we only consider the Caputo derivative by Michele Caputo and Mauro Fabrizio in [25] (which is most widely used [38] and has the same Laplace transform as the integer-order one, so it is widely used in control theory) that leads to an initial condition

© The Author(s), under exclusive license to Springer Nature Switzerland AG 2022
K. Ammari et al., *Stabilization for Some Fractional-Evolution Systems*,
SpringerBriefs in Mathematics, https://doi.org/10.1007/978-3-031-17343-1_1

which is physically meaningful [53]. This derivative possesses very interesting properties, for instance, the possibility to describe fluctuations and structures with different scales. Furthermore, this definition allows for the description of mechanical properties related to damage, fatigue, and material heterogeneities. These models are relevant, in particular, in the context of spatially disordered systems, porous media, fractal media, turbulent fluids and plasmas, biological media with traps, binding sites or macro-molecular crowding, stock price movements, etc. We refer the readers to [14, 15, 46] and the rich references therein for the motivation and description of the model. On the other hand, we refer to [21–24, 27, 35] and the rich references therein for mathematical analysis of these models.

In this book we give unified methods for the stabilization of some fractional evolution systems. More precisely, in Chap. 2, we consider the stabilization for some abstract evolution equations with a fractional damping, and in Chap. 3, we validate the abstract results of Chap. 2 on concrete examples. In Chap. 4, we study the stabilization of fractional evolution systems with memory.

Consider the following abstract dissipative wave equation:

$$\begin{cases} \partial_t^2 u(t) + A\, u(t) + B B^* \, \partial_t u = 0 \\ u(0) = u_0, \ \partial_t u(0) = u_1, \end{cases}$$

where A and B are operators mapping into some suitable Hilbert spaces. It is well known that such an energy system is decreasing over the time. Moreover, depending on the operator B or equivalently the geometric control conditions, this decay can be exponential, polynomial, or logarithmic, etc. In the first chapter of this monograph, we aim to focusing on the decay rate of the same system as above but with a fractional feedback derivative, namely the feedback is described by means of the term $B B^* \, \partial_t^\alpha u$ instead of $B B^* \, \partial_t u$. Noting that if $\mathcal{A}$ is the matrices operator corresponding to the dissipative wave equation with classical derivative, then depending on the type of the loss on the resolvent for high frequency, we obtain exponential or polynomial or logarithm stability of the semigroup associated with the operator. In this book we give an equivalent result for the system with a fractional feedback derivative. This decay rate depends on the parameter α obviously, and in Chap. 3 we give some examples to illustrate the importance of the abstract result.

If we consider $\mathcal{A}$ a generator of a C_0-semigroup of contraction the same conclusion linking loss on the resolvent and the decay rate of the semigroup associated with the operator $\mathcal{A}$ solution of the Dirichlet boundary condition

$$\begin{cases} \partial_t u(t) = \mathcal{A}\, u(t) \\ u(0) = u_0 \end{cases}$$

still true of course. The third chapter concerns with the same system but with fractional derivative instead of the classical derivative and with a memory term. By using the solution operator theory, we prove that such a system is well-posed. Next, depending on the loss type of the resolvent operator, we prove an exponential or a polynomial decay rate. Finally, by applying this result, we prove a polynomial decay rate on the energy for some general abstract integro-differential equation.

Chapter 2
Fractional Feedback Stabilization for a Class of Evolution Systems

We study the problem of stabilization for a class of evolution systems with fractional damping. After writing the equations as an augmented system, we prove in this chapter first that the problem is well-posed. Second, using LaSalle's invariance principle, we show that the energy of the system is strongly stable. Then, based on a resolvent approach, we show a lack of uniform stabilization. Next, using multiplier techniques combined with the frequency domain method, we shall give a polynomial stabilization result under some consideration on the stabilization of an auxiliary dissipating system.

2.1 Introduction

Let H be a Hilbert space equipped with the norm $\|.\|_H$, and let $A : \mathcal{D}(A) \subset H \to H$ be a self-adjoint and strictly positive operator on H. We introduce the scale of Hilbert spaces H_β, $\beta \in \mathbb{R}$, as follows: for every $\beta \geq 0$, $H_\beta = \mathcal{D}(A^\beta)$, with the norm $\|z\|_\beta = \|A^\beta z\|_H$. The space $H_{-\beta}$ is defined by duality with respect to the pivot space H as follows: $H_{-\beta} = H_\beta^*$ for $\beta > 0$. The operator A can be extended (or restricted) to each H_β, such that it becomes a bounded operator

$$A : H_\beta \to H_{\beta-1}, \quad \forall \beta \in \mathbb{R}.$$

Let a bounded linear operator $B : U \to H_{-\frac{1}{2}}$, where U is another Hilbert space which will be identified with its dual.

The system we consider here is described by

$$\begin{cases} \partial_t^2 u(t) + Au(t) + BB^* \partial_t^{\alpha,\eta} u(t) = 0, \ t > 0, \\ u(0) = u^0, \ \partial_t u(0) = u^1, \end{cases} \tag{2.1}$$

K. Ammari et al., *Stabilization for Some Fractional-Evolution Systems*, SpringerBriefs in Mathematics, https://doi.org/10.1007/978-3-031-17343-1_2

where $\partial_t^{\alpha,\eta}$ denoted the fractional derivative defined by

$$\partial_t^{\alpha,\eta} v(t) = \frac{1}{\Gamma(1-\alpha)} \int_0^t (t-s)^{-\alpha} e^{-\eta(t-s)} v'(s)\, ds, \ 0 < \alpha < 1, \ \eta \geq 0. \quad (2.2)$$

We define also the following exponentially modified fractional integro-differential operators:

$$I^{\alpha,\eta} v(t) = \frac{1}{\Gamma(\alpha)} \int_0^t (t-s)^{\alpha-1} e^{-\eta(t-s)} v(s)\, ds, \ 0 < \alpha < 1, \ \eta \geq 0.$$

With these notations, we have

$$\partial_t^{\alpha,\eta} v(t) = I^{1-\alpha,\eta} v'(t). \quad (2.3)$$

In this chapter, the fractional derivative damping force is regarded as a control force to study the properties of free damped vibration of the system, so the Caputo definition is used here.

The main result of this chapter concerns the precise asymptotic behavior of the solutions of (2.6)–(2.8). Our technique is based on a resolvent estimate.

This chapter is organized as follows. In Sect. 2.2, we reformulate problem (2.1) into an augmented system. In Sect. 2.3, we give the proper functional setting for the augmented model (2.6)–(2.8) and prove that this system is well-posed. In Sect. 2.4, we use LaSalle's invariance principle to show that the energy of the system is strongly stable. In Sect. 2.5, we prove the lack of uniform stabilization of the system (2.6)–(2.8). In Sect. 2.6, we establish a resolvent estimate which gives an explicit decay rate of the energy of the solutions of (2.6)–(2.8).

2.2 Augmented Model

In this section we reformulate (2.1) into an augmented system. Our main result is the following.

Proposition 2.2.1 *We set the constant*

$$\gamma = \frac{\sin(\alpha\pi)}{\pi},$$

and we define the function

$$p(\xi) = |\xi|^{\frac{2\alpha-1}{2}}.$$

Then the relation between the input $\mathcal{U}$ and the output $\mathcal{O}$ of the following system

$$\begin{cases} \partial_t \varphi(t, \xi) + (|\xi|^2 + \eta)\varphi(t, \xi) = p(\xi)\mathcal{U}(t) & \forall \xi \in \mathbb{R}, \ t > 0 \\ \varphi(0, \xi) = 0 & \forall \xi \in \mathbb{R} \\ \mathcal{O}(t) = \gamma \int_{\mathbb{R}} p(\xi)\varphi(t, \xi) \, d\xi, & \forall t \geq 0, \end{cases} \tag{2.4}$$

where $\mathcal{U} \in C^0([0, +\infty))$, is given by

$$\mathcal{O}(t) = I^{1-\alpha, \eta}\mathcal{U}(t). \tag{2.5}$$

Proof Solving Eq. (2.4), we obtain

$$\varphi(t, \xi) = p(\xi) \int_0^t e^{-(|\xi|^2+\eta)(t-s)}\mathcal{U}(s) \, ds.$$

It follows from the third line of (2.4) that

$$\mathcal{O}(t) = \gamma \int_0^t \mathcal{U}(s) \int_{\mathbb{R}} p(\xi)^2 e^{-(|\xi|^2+\eta)(t-s)} \, d\xi \, ds$$

$$= \frac{2 \sin(\alpha\pi)}{\pi} \int_0^t \int_0^{+\infty} \rho^{2\alpha-1} e^{-(\rho^2+\eta)(t-s)} \, d\rho \, \mathcal{U}(s) \, ds.$$

Now, using the fact that $\dfrac{1}{\Gamma(\alpha)\Gamma(1-\alpha)} = \dfrac{\sin(\alpha\pi)}{\pi}$, then, a simple change of variable leads to the relation (2.5). This completes the proof. $\quad\square$

Using now Proposition 2.2.1 and relation (2.3), system (2.1) may be recast into the following augmented system:

$$\partial_t^2 u(t) + Au(t) + \gamma B \int_{\mathbb{R}} p(\xi)\,\varphi(t, \xi) \, d\xi = 0, \ t > 0, \tag{2.6}$$

$$\partial_t \varphi(t, \xi) + (|\xi|^2 + \eta)\,\varphi(t, \xi) = p(\xi)\,B^*\partial_t u(t), \ \xi \in \mathbb{R}, \ t > 0, \tag{2.7}$$

$$u(0) = u^0, \ \partial_t u(0) = u^1, \ \varphi(0, \xi) = 0, \tag{2.8}$$

where the function $p(\xi)$ and the constant γ are given in Proposition 2.2.1.

2.3 Well-Posedness

In this section, we are interested in showing that system (2.1) is well-posed in the sense of semigroups.

Let $V = L^2(\mathbb{R}; U)$, and we set the Hilbert space $\mathcal{H} = H_{\frac{1}{2}} \times H \times V$ with inner product

$$\left\langle \begin{pmatrix} u_1 \\ v_1 \\ \varphi_1 \end{pmatrix}, \begin{pmatrix} u_2 \\ v_2 \\ \varphi_2 \end{pmatrix} \right\rangle_{\mathcal{H}} = \left\langle A^{\frac{1}{2}}u_1, A^{\frac{1}{2}}u_2 \right\rangle_H + \langle v_1, v_2 \rangle_H + \gamma \int_{\mathbb{R}} \langle \varphi_1(\xi), \varphi_2(\xi) \rangle_U \, d\xi.$$

If we put $X = \begin{pmatrix} u \\ \partial_t u \\ \varphi \end{pmatrix}$, it is clear that (2.6)–(2.8) can be written as

$$X'(t) = \mathcal{A}X(t), \quad X(0) = X_0, \tag{2.9}$$

where $X_0 = \begin{pmatrix} u_0 \\ u_1 \\ 0 \end{pmatrix}$ and $\mathcal{A} : \mathcal{D}(\mathcal{A}) \subset \mathcal{H} \to \mathcal{H}$ is defined by

$$\mathcal{A} \begin{pmatrix} u \\ v \\ \varphi \end{pmatrix} = \begin{pmatrix} v \\ -Au - \gamma \, B \int_{\mathbb{R}} p(\xi) \, \varphi(\xi) \, d\xi \\ -(|\xi|^2 + \eta)\varphi + p(\xi)B^* v \end{pmatrix}, \tag{2.10}$$

with domain

$$\mathcal{D}(\mathcal{A}) = \left\{ (u, v, \varphi) \in \mathcal{H} : v \in H_{\frac{1}{2}}, \; Au + \gamma \, B \int_{\mathbb{R}} p(\xi) \, \varphi(\xi) \, d\xi \in H, \right.$$
$$\left. |\xi|\varphi \in L^2(\mathbb{R}; U), \; -(|\xi|^2 + \eta)\varphi + p(\xi)B^* v \in L^2(\mathbb{R}; U) \right\}. \tag{2.11}$$

Our main result is given by the following theorem.

Theorem 2.3.1 *The operator $\mathcal{A}$ defined by (2.10) and (2.11) generates a C_0-semigroup of contractions $e^{t\mathcal{A}}$ in the Hilbert space $\mathcal{H}$.*

Proof To prove this result, we shall use the Lumer–Phillips theorem (see [47, Theorem 4.3]). Since for every $X = (u, v, \varphi) \in \mathcal{D}(\mathcal{A})$, we have

$$\text{Re} \, \langle \mathcal{A}X, X \rangle_{\mathcal{H}} = -\gamma \int_{\mathbb{R}} (|\xi|^2 + \eta) \|\varphi(\xi)\|_U^2 \, d\xi \leq 0,$$

then the operator $\mathcal{A}$ is dissipative.

Let $\lambda > 0$ large enough, and we prove that the operator $(\lambda I - \mathcal{A})$ is a surjection. In other words, we shall demonstrate that given any triplet $Z = (f, g, h) \in \mathcal{H}$, there is another triplet $X = (u, v, \varphi) \in \mathcal{D}(\mathcal{A})$ such that $(\lambda I - \mathcal{A})X = Z$, which can be recast as follows:

$$\begin{cases} v = \lambda u - f, \\ (\lambda^2 I + A)u = \lambda f + g - \gamma B \displaystyle\int_{\mathbb{R}} p(\xi)\varphi(\xi)\,d\xi, \\ \varphi(\xi) = \dfrac{p(\xi)}{|\xi|^2 + \eta + \lambda} B^* v + \dfrac{h(\xi)}{|\xi|^2 + \eta + \lambda}. \end{cases}$$

Since A is a nonnegative operator, then according to [59, Proposition 3.3.5], $-A$ is m-dissipative. Thus the operator $(\lambda^2 + A)$ is a bijection and we have

$$\|(\lambda^2 I + A)^{-1}\|_{\mathcal{L}(H)} \le \frac{1}{\lambda^2}.$$

Let (u_n), (v_n), and (φ_n) be three sequences defined by induction as follows:

$$\begin{cases} u_0 = (\lambda^2 I + A)^{-1}(\lambda f + g) \in H_1 \subset H_{\frac{1}{2}}, \\ v_0 = -f \in H^{\frac{1}{2}} \subset H, \\ \varphi_0(\xi) = \dfrac{h(\xi)}{|\xi|^2 + \eta + \lambda} \in V, \end{cases}$$

and

$$\begin{cases} u_{n+1} = -\gamma(\lambda^2 I + A)^{-1} B \displaystyle\int_{\mathbb{R}} p(\xi)\varphi_n(\xi)\,d\xi, \\ v_{n+1} = \lambda u_n, \\ \varphi_{n+1}(\xi) = \dfrac{p(\xi)}{|\xi|^2 + \eta + \lambda} B^* v_n. \end{cases}$$

We denote the constants C_1, C_2, and C_3 by

$$C_1 = \|f\|_{H_{-\frac{1}{2}}} + \|g\|_{H_{-\frac{1}{2}}}, \quad C_2 = \|f\|_{H_{-\frac{1}{2}}}, \quad C_3 = \left(\int_{\mathbb{R}} \frac{(1 + |\xi|)^2}{(|\xi|^2 + \eta + 1)^2}\,d\xi \right)^{\frac{1}{2}} \|h\|_V,$$

and we set the constants

$$K_1 = \gamma \|B\|_{\mathcal{L}(U, H_{-\frac{1}{2}})} \left(\int_{\mathbb{R}} \frac{p(\xi)^2}{(1 + |\xi|)^2}\,d\xi \right)^{\frac{1}{2}}$$

and

$$K_2 = \|B^*\|_{\mathcal{L}(H_{\frac{1}{2}}, U)} \left(\int_{\mathbb{R}} \left(\frac{p(\xi)(1 + |\xi|)}{|\xi|^2 + \eta + 1} \right)^2 d\xi \right)^{\frac{1}{2}},$$

from which it is clear that they are well defined.

We set the sequences

$$a_n = \lambda^{5/3} \|u_n\|_{H_{-\frac{1}{2}}}, \quad b_n = \lambda^{1/3} \|v_n\|_{H_{-\frac{1}{2}}} \text{ and } c_n = \|(1+|\xi|).\varphi_n\|_{L^2(\mathbb{R},U)}.$$

It is clear for $\lambda \geq 1$ using the Hölder inequality that

$$a_{n+1} = \lambda^{5/3} \|u_{n+1}\|_{H_{-\frac{1}{2}}} \leq K_1 \lambda^{-1/3} \|(1+|\xi|).\varphi_n\|_{L^2(\mathbb{R},U)} = K_1 \lambda^{-1/3} c_n$$

$$b_{n+1} = \lambda^{1/3} \|v_{n+1}\|_{H_{-\frac{1}{2}}} \leq \lambda^{4/3} \|u_n\|_{H_{-\frac{1}{2}}} = \lambda^{-1/3} a_n$$

$$c_{n+1} = \|(1+|\xi|).\varphi_{n+1}\|_{L^2(\mathbb{R},U)} l \leq K_2 \|v_n\|_{H_{-\frac{1}{2}}} = K_2 \lambda^{-1/3} b_n.$$

For $K = \max(1, K_1, K_2)$, we have $a_n + b_n + c_n \leq K^n \lambda^{-n/3}(a_0 + b_0 + c_0)$.

So that, for $\lambda > 0$ large enough, the two sums $\sum u_n$ and $\sum v_n$ converge uniformly in $H_{-\frac{1}{2}}$ and the sum $\sum \varphi_n$ converges uniformly in V. Therefore, by setting $u = \sum_{n=0}^{+\infty} u_n$, $v = \sum_{n=0}^{+\infty} v_n$, and $\varphi = \sum_{n=0}^{+\infty} \varphi_n$, we find

$$u = u_0 + \sum_{n=1}^{+\infty} u_n = (\lambda^2 I + A)^{-1}(\lambda f + g) - \gamma \sum_{n=1}^{+\infty} (\lambda^2 I + A)^{-1} B \int_{\mathbb{R}} p(\xi) \varphi_{n-1}(\xi) \, d\xi$$

$$= (\lambda^2 I + A)^{-1} \left((\lambda f + g) - \gamma B \int_{\mathbb{R}} p(\xi) \sum_{n=0}^{+\infty} \varphi_n(\xi) \, d\xi \right)$$

$$= (\lambda^2 I + A)^{-1}(\lambda f + g) - \gamma(\lambda^2 I + A)^{-1} B \int_{\mathbb{R}} p(\xi) \varphi(\xi) \, d\xi.$$

Since $\varphi \in V$, we follow that $u \in H_{\frac{1}{2}}$, and we have

$$(\lambda^2 I + A)u = (\lambda f + g) - \gamma B \int_{\mathbb{R}} p(\xi) \varphi(\xi) \, d\xi.$$

By the same way, we have

$$v = \sum_{n=0}^{+\infty} v_n = v_0 + \sum_{n=1}^{+\infty} v_n = f + \lambda \sum_{n=1}^{+\infty} u_{n-1} = \lambda u + f$$

and also

$$\varphi(\xi) = \sum_{n=0}^{+\infty} \varphi_n(\xi) = \varphi_0(\xi) + \sum_{n=1}^{+\infty} \varphi_n(\xi) = \frac{h(\xi)}{|\xi|^2 + \eta + \lambda} + \frac{p(\xi)}{|\xi|^2 + \eta + \lambda} B^* \sum_{n=1}^{+\infty} v_{n-1}$$

$$= \frac{h(\xi)}{|\xi|^2 + \eta + \lambda} + \frac{p(\xi)}{|\xi|^2 + \eta + \lambda} B^* v.$$

This proves $v \in H_{\frac{1}{2}}$ and $|\xi| \varphi, \varphi \in L^2(\mathbb{R}; U)$. Finally, it is clear that

$$Au + \gamma B \int_{\mathbb{R}} p(\xi) \varphi(\xi) \, d\xi \in H \text{ and } -(|\xi|^2 + \eta)\varphi + p(\xi) B^* v \in L^2(\mathbb{R}; U).$$

Hence, we proved that the operator $(\mathcal{A} - \lambda I)$ is onto. This completes the proof. □

As a consequence of Theorem 2.3.1, the system (2.6)–(2.8) is well-posed in the energy space $\mathcal{H}$, and we have the following proposition.

Proposition 2.3.1 *For* $(u^0, u^1, 0) \in \mathcal{H}$, *the problem* (2.6)–(2.8) *admits a unique solution*

$$(u, \partial_t u, \varphi) \in C([0, +\infty); \mathcal{H}),$$

and for $(u^0, u^1, 0) \in \mathcal{D}(\mathcal{A})$, *the problem* (2.6)–(2.8) *admits a unique solution*

$$(u, \partial_t u, \varphi) \in C([0, +\infty); \mathcal{D}(\mathcal{A})) \cap C^1([0, +\infty); \mathcal{H}).$$

Moreover, from the density of $\mathcal{D}(\mathcal{A})$ *in* $\mathcal{H}$, *the energy of* $(u(t), \varphi(t))$ *at time* $t \geq 0$ *given by*

$$E(t) = \frac{1}{2} \left(\|u(t)\|_{H_{\frac{1}{2}}}^2 + \|\partial_t u(t)\|_H^2 + \gamma \int_{\mathbb{R}} \|\varphi(t, \xi)\|_U^2 \, d\xi \right)$$

decays as follows:

$$\frac{dE}{dt}(t) = -\gamma \int_{\mathbb{R}} \left(|\xi|^2 + \eta \right) \|\varphi(t, \xi)\|_U^2 \, d\xi, \ \forall \, t \geq 0. \tag{2.12}$$

Proof Noting that the regularity of the solution of the problem (2.6)–(2.8) is a consequence of the semigroup properties. We have just to prove (2.12). We set

$$E_1(t) = \frac{1}{2} \left(\|u(t)\|_{H_{\frac{1}{2}}}^2 + \|\partial_t u(t)\|_H^2 \right) \quad \text{and} \quad E_2(t) = \frac{\gamma}{2} \left(\int_{\mathbb{R}} \|\varphi(t, \xi)\|_U^2 \, d\xi \right).$$

A straightforward calculation gives

$$\frac{dE_1}{dt}(t) = -\gamma \operatorname{Re}\left\langle \int_{\mathbb{R}} p(\xi)\varphi(t,\xi)\, d\xi, B^* \partial_t u(t)\right\rangle_U$$

and

$$\frac{dE_2}{dt}(t) = \gamma \operatorname{Re}\left\langle \int_{\mathbb{R}} p(\xi)\varphi(t,\xi)\, d\xi, B^* \partial_t u(t)\right\rangle_U - \gamma \int_{\mathbb{R}} (|\xi|^2 + \eta).\|\varphi(t,\xi)\|_U^2\, d\xi.$$

Since $E(t) = E_1(t) + E_2(t)$, then (2.12) holds and this completes the proof.   □

2.4 Strong Stabilization

In this section, we prove that the solutions of system (2.9) converge asymptotically to zero. To achieve this result, we shall make use of LaSalle's invariance principle extended to infinite-dimensional systems [61]. According to this principle, all solutions of (2.9) will asymptotically tend to the maximal invariant subset of the set

$$\mathcal{I} = \left\{ X \in \mathcal{H} : \frac{dE}{dt}(t) = 0 \right\},$$

provided that these solutions are pre-compact in $\mathcal{H}$.

Lemma 2.4.1 *Let*

$$\mathcal{E}(t) = \frac{1}{2}\left(\|\partial_t u\|_{H_{\frac{1}{2}}}^2 + \|\partial_t^2 u\|_H^2 + \gamma \int_{\mathbb{R}} \|\partial_t \varphi(\xi)\|_U^2\, d\xi \right).$$

Then the function $t \longmapsto \mathcal{E}(t)$ is non-increasing along solutions of the system (2.9) with initial data in $\mathcal{D}(\mathcal{A}^2)$. In particular, we have

$$\frac{d\mathcal{E}}{dt}(t) = -\gamma \int_{\mathbb{R}} (|\xi|^2 + \eta).\|\partial_t \varphi(t,\xi)\|_U^2\, d\xi. \tag{2.13}$$

Proof If $X_0 \in \mathcal{D}(\mathcal{A}^2)$, then the $X(t) = e^{t\mathcal{A}}X_0$ is a solution of (2.9) with the following regularity:

$$X(t) = \begin{pmatrix} u(t) \\ \partial_t u(t) \\ \varphi(t) \end{pmatrix} \in \mathcal{C}([0, +\infty[, \mathcal{D}(\mathcal{A}^2)) \cap \mathcal{C}^1([0, +\infty[, \mathcal{D}(\mathcal{A}))$$

with $\dot{X}(t) = \begin{pmatrix} \partial_t u(t) \\ \partial_t^2 u(t) \\ \partial_t \varphi(t,\xi) \end{pmatrix} = \mathcal{A} X(t) = \mathcal{A} e^{t\mathcal{A}} X_0 = e^{t\mathcal{A}} \mathcal{A} X_0$. And since $\mathcal{A} X_0 \in$
$\mathcal{D}(\mathcal{A})$, then

$$\dot{X}(t) \in \mathcal{C}([0,+\infty[, \mathcal{D}(\mathcal{A})) \cap \mathcal{C}^1([0,+\infty[, \mathcal{H}),$$

then by setting

$$\mathcal{E}_1(t) = \frac{1}{2} \left(\|\partial_t u\|_{H_{\frac{1}{2}}}^2 + \|\partial_t^2 u\|_H^2 \right) \quad \text{and} \quad \mathcal{E}_2(t) = \frac{\gamma}{2} \left(\int_{\mathbb{R}} \|\partial_t \varphi(\xi)\|_U^2 \, d\xi \right),$$

we have

$$\frac{d\mathcal{E}_1}{dt}(t) = -\gamma \mathrm{Re} \left\langle \int_{\mathbb{R}} p(\xi) \partial_t \varphi(t,\xi) \, d\xi, B^* \partial_t^2 u \right\rangle_U$$

and

$$\frac{d\mathcal{E}_2}{dt}(t) = \gamma \mathrm{Re} \left\langle \int_{\mathbb{R}} p(\xi) \partial_t \varphi(t,\xi) \, d\xi, B^* \partial_t^2 u \right\rangle_U - \gamma \int_{\mathbb{R}} (|\xi|^2 + \eta) . \|\partial_t \varphi(t,\xi)\|_U^2 \, d\xi.$$

So that, by summing the last two expressions, we obtain (2.13), and consequently the non-increasing property of $\mathcal{E}(t)$ holds. This completes the proof. $\qquad\square$

Lemma 2.4.2 *We assume that the only classical solution $u(t)$ (i.e., for all $t \geq 0$, $(u(t), \partial_t u(t)) \in H_{\frac{1}{2}} \times H$) of the following system:*

$$\begin{cases} \partial_t^2 u(t) + Au(t) = 0 \\ B^* \partial_t u(t) = 0 \end{cases} \tag{2.14}$$

is the trivial one, and then the only classical solution (i.e. for all $t \geq 0$, $X(t) \in \mathcal{H}$) of (2.9) in the subspace $\mathcal{I}$ is the zero solution.

Proof Let $X = (u, v, \varphi) \in \mathcal{I}$ be a classical solution of (2.9). Then, from (2.12), we have

$$\int_{\mathbb{R}} \left(|\xi|^2 + \eta \right) \|\varphi(t,\xi)\|_U^2 \, d\xi = 0,$$

which implies that

$$\varphi(t,\xi) \equiv 0 \quad \text{in } L^2(\mathbb{R}; U) \ \forall t \geq 0. \tag{2.15}$$

By using (2.15), it is clear that system (2.9) reduces to the system (2.14). Then by the assumption made in this lemma, we deduce that $u(t) \equiv 0$ for all $t \geq 0$. This completes the proof.   □

Proposition 2.4.1 *Let* $X_0 = (u_0, v_0, \varphi_0) \in \mathcal{D}(\mathcal{A}^2)$, *and then the trajectory of* $\varphi(t)$, *the third component of the solution of (2.9), is pre-compact in* $L^2(\mathbb{R}; U)$.

Proof Since, for $X_0 \in \mathcal{D}(\mathcal{A}^2)$, $\varphi(t)$ is continuous mapping from $[0, +\infty[$ into $L^2(\mathbb{R}, U)$, it is therefore sufficient to show that

$$\int_{\mathbb{R}} \|\varphi(t, \xi)\|_U^2 \, d\xi \longrightarrow 0 \quad \text{as} \quad t \nearrow +\infty.$$

From (2.12) and (2.13) together with the fact that both $E(t)$ and $\mathcal{E}(t)$ are non-increasing functions, we follow

$$\int_0^{+\infty} \int_{\mathbb{R}} (|\xi|^2 + \eta) \|\varphi(t, \xi)\|_U^2 \, d\xi \, dt < +\infty, \tag{2.16}$$

and

$$\int_0^{+\infty} \int_{\mathbb{R}} (|\xi|^2 + \eta) \|\partial_t \varphi(t, \xi)\|_U^2 \, d\xi \, dt < +\infty. \tag{2.17}$$

The remainder of the proof will be divided into two cases.

  Case 1: $\eta \neq 0$. Here we get immediately from (2.16) and (2.17) the following relations:

$$\int_0^{+\infty} \int_{\mathbb{R}} \|\varphi(t, \xi)\|_U^2 \, d\xi \, dt < +\infty \tag{2.18}$$

and

$$\int_0^{+\infty} \int_{\mathbb{R}} \|\partial_t \varphi(t, \xi)\|_U^2 \, d\xi \, dt < +\infty. \tag{2.19}$$

By using theses relations together with the well-know inequality $2 \operatorname{Re}\langle X, Y \rangle \leq \|X\|^2 + \|Y\|^2$ for all X, Y, we obtain

$$\left| \int_{\mathbb{R}} \|\varphi(t, \xi)\|_U^2 \, d\xi - \int_{\mathbb{R}} \|\varphi(s, \xi)\|_U^2 \, d\xi \right| = 2 \left| \operatorname{Re} \left(\int_s^t \int_{\mathbb{R}} \langle \partial_t \varphi(t, \xi), \varphi(t, \xi) \rangle_U \, d\xi \, dt \right) \right|$$

$$\leq \int_s^t \int_{\mathbb{R}} \|\partial_t \varphi(\xi, t)\|_U^2 + \|\varphi(\xi, t)\|_U^2 \, d\xi \, dt,$$

and then we easily see from (2.18) and (2.19) that

$$\lim_{t \to +\infty} \int_{\mathbb{R}} \|\varphi(t, \xi)\|_U^2 \, d\xi \quad \text{exist and finite.} \tag{2.20}$$

But then (2.18) and (2.20) imply that

$$\lim_{t \to +\infty} \int_{\mathbb{R}} \|\varphi(t, \xi)\|_U^2 \, d\xi = 0.$$

Case 2: $\eta = 0$. In this case (2.16) and (2.17) reduce to

$$\int_0^{+\infty} \int_{\mathbb{R}} |\xi|^2 \|\varphi(t, \xi)\|_U^2 \, d\xi \, dt < +\infty \tag{2.21}$$

and

$$\int_0^{+\infty} \int_{\mathbb{R}} |\xi|^2 \|\partial_t \varphi(t, \xi)\|_U^2 \, d\xi \, dt < +\infty. \tag{2.22}$$

Again, by using the inequality $2 \operatorname{Re}\langle X, Y \rangle \leq \|X\|^2 + \|Y\|^2$, we have

$$\lim_{t \to +\infty} \int_{\mathbb{R}} |\xi|^2 \|\varphi(t, \xi)\|_U^2 \, d\xi \quad \text{exist and finite.}$$

Thus, (2.21) implies that

$$\lim_{t \to +\infty} \int_{\mathbb{R}} |\xi|^2 \|\varphi(t, \xi)\|_U^2 \, d\xi = 0. \tag{2.23}$$

Therefore, in view of (2.23), it is clear that $\int_{\mathbb{R}} \|\varphi(t, \xi)\|_U^2 \, d\xi$ will tend to zero as t goes to $+\infty$, if

$$\lim_{t \to +\infty} \int_{B(0,1)} \|\varphi(t, \xi)\|_U^2 \, d\xi = 0, \tag{2.24}$$

where $B(0, 1)$ is the unit ball in $\mathbb{R}$. Next, we prove (2.24) by using the dominated convergence theorem whose conditions of applicability, in the case at hand, are established below:

$(*)$ By applying Fubini's theorem to both inequalities (2.21) and (2.22), we have

$$\int_0^{+\infty} |\xi|^2 \|\varphi(\xi, t)\|_U^2 \, dt < +\infty \text{ a.e. } \xi \in B(0, 1)$$

and

$$\int_0^{+\infty} |\xi|^2 \|\partial_t \varphi(\xi, t)\|_U^2 \, dt < +\infty \quad \text{a.e. } \xi \in B(0, 1).$$

So that, by the same argument that led us to (2.23), we may conclude that

$$\lim_{t \to +\infty} |\xi|^2 \|\varphi(t, \xi)\|_U^2 = 0 \qquad \text{a.e. } \xi \in B(0, 1).$$

Hence, we obtain

$$\lim_{t \to +\infty} \|\varphi(t, \xi)\|_U^2 = 0 \qquad \text{a.e. } \xi \in B(0, 1). \tag{2.25}$$

($*$) Now, solving (2.7), we have

$$\varphi(t, \xi) = \varphi_0(\xi) e^{-|\xi|^2 t} + p(\xi) B^* \int_0^t \partial_t u(s) e^{-|\xi|^2 (t-s)} \, ds. \tag{2.26}$$

So that, by applying integration by parts, to the integral in the right-hand side of (2.26), we get

$$\varphi(t, \xi) = \varphi_0(\xi) e^{-|\xi|^2 t}$$
$$+ p(\xi) B^* [u(t) - u(0) e^{-|\xi|^2 t}] - |\xi|^2 p(\xi) B^* \int_0^t u(s) e^{-|\xi|^2 (t-s)} \, ds.$$

Hence, one gets

$$\|\varphi(t, \xi)\|_U \leq \|\varphi_0(\xi)\|_U + p(\xi) \|B^*\|_{\mathcal{L}(H_{\frac{1}{2}}, U)}$$
$$\times \left[\|u(t)\|_{H_{\frac{1}{2}}} + \|u(0)\|_{H_{\frac{1}{2}}} + |\xi|^2 \int_0^t \|u(s)\|_{H_{\frac{1}{2}}} e^{-|\xi|^2 (t-s)} \, ds \right].$$

Also, by (2.12), we can bound $\|u(t)\|_{H_{\frac{1}{2}}}^2 \leq E(0)$ and we obtain

$$\|\varphi(t, \xi)\|_U^2 \leq \|\varphi_0(\xi)\|_U^2 + p(\xi)^2 \|B^*\|_{\mathcal{L}(H_{\frac{1}{2}}, U)}^2 \left[2E(0) + E(0)|\xi|^2 (1 - e^{-|\xi| t}) \right]. \tag{2.27}$$

Since the right-hand side of (2.27) is in $L_\xi^1(B(0, 1))$, therefore by combining (2.25) and (2.27) through the dominated convergence theorem, we get (2.24) the desired result. □

Proposition 2.4.2 *We suppose that B is bounded, which means* $B : U \longrightarrow H$ *is bounded, and that the embedding* $H_{\frac{1}{2}} \hookrightarrow H$ *is compact. Let* $X_0 = (u_0, v_0, \varphi_0) \in \mathcal{D}(\mathcal{A}^2)$, *and then the trajectory of the pair* $(u(t), v(t))$ *of the solution of the system* (2.9) *is pre-compact in* $H_{\frac{1}{2}} \times H$.

Proof Note that if $X_0 = (u_0, v_0, \varphi_0) \in \mathcal{D}(\mathcal{A}^2)$, then $(u(t), v(t)) \in H_1 \times H_{\frac{1}{2}}$. Since that, in view of the assumption made in this proposition, it is clear that to prove this result we have just to prove that the quantity $\|u(t)\|_{H_1}^2 + \|v(t)\|_{H_{\frac{1}{2}}}$ is bounded. We solve the differential equation (2.7), and we get

$$\varphi(t, \xi) = \varphi_0(\xi) e^{-(|\xi|^2 + \eta)t} + p(\xi) B^* \int_0^t \partial_t u(s) e^{-(|\xi|^2 + \eta)(t-s)} \, ds$$

$$= \varphi_0(\xi) e^{-(|\xi|^2 + \eta)t} + p(\xi) B^* \int_0^t \partial_t u(t - s) e^{-(|\xi|^2 + \eta)s} \, ds. \tag{2.28}$$

Using the differential equation (2.6) and Fubini's theorem and taking account of (2.28) and the fact that $\mathcal{E}(t)$ is bounded by $\mathcal{E}(0)$, we have

$$\|u(t)\|_{H_1}^2 + \|v(t)\|_{H_{\frac{1}{2}}}^2 = \|Au\|_H^2 + \|\partial_t u\|_{H_{\frac{1}{2}}}^2$$

$$\leq C \left(\mathcal{E}(0) + \|B\|_{\mathcal{L}(U,H)}^2 \left\| \int_{\mathbb{R}} p(\xi) \varphi_0(\xi) e^{-(|\xi|^2 + \eta)t} \, d\xi \right\|_U^2 \right.$$

$$\left. + \|BB^*\|_{\mathcal{L}(H)}^2 \left\| \int_{\mathbb{R}} p(\xi)^2 \int_0^t \partial_t u(t - s) e^{-(|\xi|^2 + \eta)s} \, ds \, d\xi \right\|_H^2 \right)$$

$$\leq C \left(\mathcal{E}(0) + \|B\|_{\mathcal{L}(U,H)}^2 \int_0^{+\infty} \frac{\rho^{2\alpha - 1}}{(1 + \rho^2)} \, d\rho . \int_{\mathbb{R}} (1 + |\xi|^2) \|\varphi_0(\xi)\|_U^2 \, d\xi \right.$$

$$\left. + \|BB^*\|_{\mathcal{L}(H)}^2 \left\| \int_0^t \int_0^{+\infty} \rho^{2\alpha - 1} \partial_t u(t - s) e^{-(\rho^2 + \eta)s} \, d\rho \, ds \right\|_H^2 \right). \tag{2.29}$$

Now we set

$$I = \left\| \int_0^t \int_0^{+\infty} \rho^{2\alpha - 1} \partial_t u(t - s) e^{-(\rho^2 + \eta)s} \, d\rho \, ds \right\|_H^2,$$

and to establish our result, it is clear that we have just to prove that I is bounded. To do so, we distinguish two cases.

Case 1: $\eta \neq 0$. Using again Fubini's theorem and the fact that $\mathcal{E}(t)$ is a non-increasing function, we obtain

$$I \leq 2\mathcal{E}(0) \left(\int_0^{+\infty} \int_0^t \rho^{2\alpha-1} e^{-(\rho^2+\eta)s} \, ds \, d\rho \right)^2$$

$$= 2\mathcal{E}(0) \left(\int_0^{+\infty} \frac{\rho^{2\alpha-1}}{\rho^2+\eta} \left(1 - e^{-(\rho^2+\eta)t} \right) d\rho \right)^2$$

$$\leq 4\mathcal{E}(0) \left(\int_0^{+\infty} \frac{\rho^{2\alpha-1}}{\rho^2+\eta} \, d\rho \right)^2 < +\infty$$

which proves that I is bounded.

Case 2: $\eta = 0$. It is clear that according to the first case that the problem of the boundedness of I is reduced to the boundedness of the following integral:

$$I_0 = \left\| \int_1^t \int_0^1 \rho^{2\alpha-1} \partial_t u(t-s) e^{-\rho^2 s} \, d\rho \, ds \right\|_H^2,$$

where we can suppose that $t \geq 1$. Integrating by parts with respect to the s variable and using again the fact that $E(t)$ is a non-increasing function, we have

$$I_0 \leq 2 \left(\left\| \int_0^1 \rho^{2\alpha-1} \left(e^{-t\rho^2} u(0) - e^{-\rho^2} u(t-1) \right) d\rho \right\|_H^2 \right.$$

$$+ \left\| \int_0^1 \rho^{2\alpha+1} \int_1^t e^{-s\rho^2} u(t-s) \, ds \, d\rho \right\|_H^2 \right)$$

$$\leq C \left(E(0) \left(\int_0^1 \rho^{2\alpha-1} d\rho \right)^2 + E(0) \left(\int_0^1 \rho^{2\alpha-1} (e^{-\rho^2} - e^{-t\rho^2}) \, d\rho \right)^2 \right)$$

$$\leq C E(0).$$

This proves the expected estimate and ends the proof.   □

Theorem 2.4.1 *We assume that the embedding $H_{\frac{1}{2}} \hookrightarrow H$ is compact and that the unique classical solution of system (2.14) is the trivial one, then the semigroup e^{tA} is strongly stable, and it means that for any initial data $X_0 \in \mathcal{H}$,*

$$\|e^{tA} X_0\|_{\mathcal{H}} \longrightarrow 0 \quad as \quad t \longrightarrow +\infty.$$

Proof For $X_0 \in \mathcal{D}(\mathcal{A}^2)$, the theorem is a direct consequence of Lemma 2.4.2, Propositions 2.4.1 and 2.4.2 and LaSalle's invariance principle. Finally, since $\mathcal{D}(\mathcal{A}^2)$ is dense in $\mathcal{H}$, this result carries over all $X_0 \in \mathcal{H}$.   □

2.5 Lack of Uniform Stabilization

In this section we shall prove that system is not uniformly exponentially stable.

Lemma 2.5.1 *Let $\omega \in \mathbb{R}^*$. For any fixed $\eta > 0$ and $0 < \alpha < 1$, we have*

$$\int_0^{+\infty} \frac{\rho^{2\alpha-1}}{\rho^2 + \eta + i\omega} \, d\rho = \begin{cases} \dfrac{-\pi(1 + e^{-2i\alpha\pi})}{2(\eta^2 + \omega^2)^{\frac{1-\alpha}{2}} \sin(2\alpha\pi)} e^{2i(\alpha-1)\theta} & \text{if } \alpha \neq \dfrac{1}{2} \\[4mm] \dfrac{-i\pi}{2(\eta^2 + \omega^2)^{\frac{1}{4}} e^{i\theta}} & \text{if } \alpha = \dfrac{1}{2}, \end{cases}$$

(2.30)

where we have denoted $\theta = \arccos\left(-\dfrac{\sqrt{\dfrac{\sqrt{\eta^2 + \omega^2} - \eta}{2}}}{(\eta^2 + \omega^2)^{\frac{1}{4}}}\right).$

Proof The two cases are proven as follows:

Case 1: $\alpha \neq \dfrac{1}{2}$. In this case the integral can be evaluated using the method of residues; by integrating along the positive oriented contour drawn in Fig. 2.1. We set the function

$$f(z) = \frac{z^{2\alpha-1}}{z^2 + \eta + i\omega}, \qquad \forall z \in \mathbb{C} \setminus \mathbb{R}_-,$$

Fig. 2.1 Contour for evaluating the integral $\displaystyle\int_0^{+\infty} \frac{\rho^{2\alpha-1}}{\rho^2 + \eta + i\omega} \, d\rho$

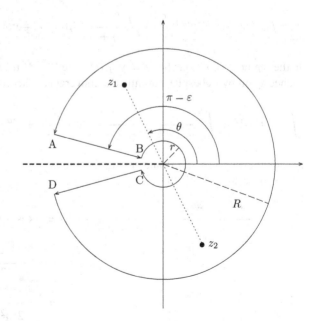

whose poles are $z_1 = (\eta^2 + \omega^2)^{\frac{1}{4}} e^{i\theta}$, $z_2 = (\eta^2 + \omega^2)^{\frac{1}{4}} e^{i(\theta - \pi)}$, and eventually $z_0 = 0$ (see Fig. 2.1). Clearly, we have

$$|zf(z)| \leq \frac{|z|^{2\alpha}}{\left| |z|^2 - (\eta^2 + \omega^2)^{\frac{1}{2}} \right|}, \tag{2.31}$$

which implies that

$$\lim_{z \to 0} zf(z) = 0, \qquad \lim_{|z| \to +\infty} zf(z) = 0.$$

Then, by Jordan's lemma, it follows

$$\lim_{r \to 0} \int_{\gamma_r} f(z) \, dz = 0 \tag{2.32}$$

and

$$\lim_{R \to +\infty} \int_{\gamma_R} f(z) \, dz = 0, \tag{2.33}$$

where $\gamma_r = re^{-it}$ and $\gamma_R = re^{it}$ for $t \in [-\pi + \varepsilon, \pi - \varepsilon]$ (see Fig. 2.1).

On the segment $[AB]$, one has $z = \gamma_{AB}(t) = [(1-t)R + rt] e^{i(\pi - \varepsilon)}$ for $t \in [0, 1]$ (see Fig. 2.1), whence by Lebesgue dominated convergence theorem, we have

$$\int_{\gamma_{AB}} f(z) \, dz = e^{i(\varepsilon + 2\alpha\pi)} \int_r^R \frac{\rho^{2\alpha - 1}}{\rho^2 + \eta + i\omega} \, d\rho \xrightarrow[\varepsilon \to 0]{} -e^{2i\alpha\pi} \int_r^R \frac{\rho^{2\alpha - 1}}{\rho^2 + \eta + i\omega} \, d\rho. \tag{2.34}$$

On the segment $[CD]$ one has $z = \gamma_{CD}(t) = te^{i(-\pi + \varepsilon)}$ for $t \in [r, R]$ (see Fig. 2.1), whence again by Lebesgue's dominated convergence theorem we have

$$\int_{\gamma_{CD}} f(z) \, dz = \int_r^R \frac{\rho^{2\alpha - 1} e^{2i\alpha(\varepsilon - \pi)}}{\rho^2 e^{2i(\varepsilon - \pi)} + \eta + i\omega} \, d\rho \xrightarrow[\varepsilon \to 0]{} e^{-2i\alpha\pi} \int_r^R \frac{\rho^{2\alpha - 1}}{\rho^2 + \eta + i\omega} \, d\rho. \tag{2.35}$$

By summing (2.32)–(2.35) and taking the limits as $r \searrow 0$ and $R \nearrow +\infty$, the method of residues leads to

$$\frac{e^{-2i\alpha\pi} - e^{2i\alpha\pi}}{2i\pi} \int_0^{+\infty} \frac{\rho^{2\alpha - 1}}{\rho^2 + \eta + i\omega} \, d\rho = \operatorname*{Res}_{z = z_1, z_2} [f(z)]$$

$$= \frac{z_1^{2\alpha - 2} + z_2^{2\alpha - 2}}{2}$$

$$= \frac{e^{2i(\alpha - 1)\theta} [1 + e^{-2i\alpha\pi}]}{2(\eta^2 + \omega^2)^{\frac{1-\alpha}{2}}},$$

which leads to the second line of (2.30).

Case 2: $\alpha = \dfrac{1}{2}$. Since z_1 and z_2 are the unique poles of f, then we can write

$$\int_0^{+\infty} \frac{d\rho}{\rho^2 + \eta + i\omega} = \frac{1}{2z_1} \int_0^{+\infty} \frac{\rho - \bar{z}_1}{\rho^2 - 2\mathrm{Re}(z_1)\rho + |z_1|^2} - \frac{\rho - \bar{z}_2}{\rho^2 - 2\mathrm{Re}(z_2)\rho + |z_2|^2}\, d\rho$$

$$= \frac{1}{2z_1} \int_0^{+\infty} \frac{\rho - \bar{z}_1}{(\rho - \mathrm{Re}(z_1))^2 + \mathrm{Im}(z_1)^2} - \frac{\rho - \bar{z}_2}{(\rho - \mathrm{Re}(z_2))^2 + \mathrm{Im}(z_2)^2}\, d\rho.$$

A straightforward calculation leads to

$$\int_0^{+\infty} \frac{d\rho}{\rho^2 + \eta + i\omega} = \frac{-i\pi}{2z_1},$$

which leads to the first line of (2.30). And this completes the proof. $\square$

Let us assume that H is an infinite-dimensional Hilbert space such that the imbedding $H_{\frac{1}{2}} \hookrightarrow H$ is compact. Since A is a positive operator with compact resolvent, then there exists a sequence of eigenvalues $i\omega_n$ corresponding to the orthonormal base of the eigenfunctions $\phi_n = \begin{pmatrix} u_n \\ i\omega_n \\ u_n \end{pmatrix}$ of the operator $\mathcal{A}_0 = \begin{pmatrix} 0 & I \\ -A & 0 \end{pmatrix}$ such that $\displaystyle\lim_{n\to+\infty} |\omega_n| = +\infty$ where $u_n \in H_{\frac{1}{2}}$.

Theorem 2.5.1 *Under the above assumptions, we have*

(1) If $B \in \mathcal{L}(U, H)$, the semigroup $e^{t\mathcal{A}}$ is not exponentially stable in the Hilbert space $\mathcal{H}$.

(2) If $B \in \mathcal{L}(U, H_{-\frac{1}{2}})$, the semigroup $e^{t\mathcal{A}}$ is not exponentially stable in the Hilbert space $\mathcal{H}$, provided that one of the following statements holds:

   (i) For some $n \in \mathbb{N}$, $B^* u_n = 0$. (In this case the semigroup $e^{t\mathcal{A}}$ is not even strongly stable and the energy is conserved for some initial data and some eigenvalues lies on the imaginary axis.)

   (ii) For all $n \in \mathbb{N}$, $B^* u_n \neq 0$ and there exists a subsequence (u_{n_k}) verifying $BB^* u_{n_k} \in H$ and $\|BB^* u_{n_k}\|_H \leq C$ for some $C > 0$ and for every $k \in \mathbb{N}$.

Proof To prove this theorem, we shall use the frequency theorem method. We recall that a bounded C_0-semigroup generated by an operator $\mathcal{A}$ is exponentially stable if and only if $i\mathbb{R} \cap \sigma(\mathcal{A}) = \varnothing$, and it satisfies the following identity:

$$\limsup_{\omega \in \mathbb{R}, |\omega| \to +\infty} \|(i\omega I - \mathcal{A})^{-1}\|_{\mathcal{L}(\mathcal{H})} < +\infty.$$

We distinguish now two cases.

Case 1: $B^* u_n = 0$ for some $n \in \mathbb{N}$. It is clear in this case that for such an $n \in \mathbb{N}$ we have $X_n = \begin{pmatrix} \frac{u_n}{i\omega_n} \\ u_n \\ 0 \end{pmatrix} \in \mathcal{D}(\mathcal{A})$ and $(i\omega_n I - \mathcal{A})X_n = 0$, which proves that X_n is an eigenfunction corresponding to the eigenvalue $i\omega_n$. Thus, the semigroup $e^{t\mathcal{A}}$ is not uniformly stable.

Case 2: $B^* u_n \neq 0$ for all $n \in \mathbb{N}$. In this part we shall prove a more general result than given in the theorem. In fact, we will show that the following resolvent estimate:

$$\limsup_{\omega \in \mathbb{R}, |\omega| \to +\infty} \|\omega^{\alpha - 1 + \varepsilon}(i\omega I - \mathcal{A})^{-1}\|_{\mathcal{L}(\mathcal{H})} < +\infty \tag{2.36}$$

is not even satisfied, for $\varepsilon > 0$ small.

Let $\varphi_n(\xi) = \dfrac{p(\xi)}{|\xi|^2 + \eta + i\omega_n} B^* u_n$ and $X_n = \begin{pmatrix} \frac{u_n}{i\omega_n} \\ u_n \\ \varphi_n \end{pmatrix}$. It is clear that the integrals

$$\int_{\mathbb{R}} \frac{p(\xi)^2}{|\xi|^2 + \eta + i\omega_n}\, d\xi \quad \text{and} \quad \int_{\mathbb{R}} \frac{|\xi|^2 p(\xi)^2}{(|\xi|^2 + \eta)^2 + \omega_n^2}\, d\xi$$

are well defined, then $|\xi| \varphi_n \in L^2(\mathbb{R}; U)$, and by the assumption made in this theorem, we have $u_n + \gamma\, B \displaystyle\int_{\mathbb{R}} p(\xi)\varphi_n(\xi)\, d\xi \in H$. Besides, since we have

$$\int_{\mathbb{R}} \|p(\xi)B^* u_n - (|\xi|^2 + \eta)\varphi_n\|_U^2\, d\xi = \omega_n^2 \|B^* u_n\|_U^2 \int_{\mathbb{R}} \frac{p(\xi)^2}{(|\xi|^2 + \eta)^2 + \omega_n^2}\, d\xi,$$

then $p(\xi)B^* u_n - (|\xi|^2 + \eta)\varphi_n \in L^2(\mathbb{R}; U)$, and this shows that $X_n \in \mathcal{D}(\mathcal{A})$.

We set now $Y_n = \begin{pmatrix} f_n \\ g_n \\ h_n \end{pmatrix} \in \mathcal{H}$ such that

$$(i\omega_n I - \mathcal{A})X_n = Y_n.$$

Since we have $f_n = h_n = 0$ and

$$g_n = \gamma\, B B^* u_n \int_{\mathbb{R}} \frac{p(\xi)^2}{|\xi|^2 + \eta + i\omega_n}\, d\xi.$$

we set now

$$
\kappa_n = \begin{cases} e^{-i\theta_n} & \text{if } \alpha = \dfrac{1}{2} \\[2mm] -\dfrac{(1+e^{-2i\alpha\pi})}{2\cos(\alpha\pi)}e^{2i(\alpha-1)\theta_n} & \text{if } \alpha \neq \dfrac{1}{2}, \end{cases}
$$

where $\theta_n = \arccos\left(-\dfrac{\sqrt{\frac{\sqrt{\eta^2+\omega_n^2}-\eta}{2}}}{(\eta^2+\omega_n^2)^{\frac{1}{4}}}\right)$. According to Lemma 2.5.1, the function g_n

can be written as follows:

$$
g_n = \frac{\gamma\kappa_n}{(\eta_n^2+\omega_n^2)^{\frac{1-\alpha}{2}}}BB^*u_n.
$$

Then, using the assumption made on the boundedness of the sequence $(\|BB^*u_n\|_H)$, we follow that

$$
\omega_n^{1-\alpha-\varepsilon}\|g_n\|_H \leq \frac{C\omega_n^{1-\alpha-\varepsilon}}{(\eta^2+\omega_n^2)^{\frac{1-\alpha}{2}}}\|BB^*u_n\|_H \longrightarrow 0 \text{ as } n \nearrow +\infty.
$$

Hence, by assuming that the imaginary axis is a subset of the resolvent set, we follow

$$
\begin{aligned}
\limsup_{\omega\in\mathbb{R},|\omega|\to+\infty}\ \|\omega^{\alpha-1+\varepsilon}(i\omega I-\mathcal{A})^{-1}\|_{\mathcal{L}(\mathcal{H})} &\geq \sup_{n\in\mathbb{N}}\|\omega_n^{\alpha-1+\varepsilon}(i\omega_n I-\mathcal{A})^{-1}\|_{\mathcal{L}(\mathcal{H})} \\[2mm]
&\geq \sup_{n\in\mathbb{N}}\omega_n^{\alpha-1+\varepsilon}\frac{\|(i\omega_n I-\mathcal{A})^{-1}(Y_n)\|_{\mathcal{H}}}{\|Y_n\|_{\mathcal{H}}} \\[2mm]
&\geq \lim_{n\to+\infty}\omega_n^{\alpha-1+\varepsilon}\frac{\|X_n\|_{\mathcal{H}}}{\|Y_n\|_{\mathcal{H}}} \\[2mm]
&\geq \lim_{n\to+\infty}\frac{\omega_n^{\alpha-1+\varepsilon}}{\|g_n\|_H} = +\infty.
\end{aligned}
$$

Thus, (2.36) is not satisfied. This proves that the semigroup $e^{t\mathcal{A}}$ is not exponentially stable. $\square$

Remark 2.5.1 In the infinite-dimensional case, provided the compactness of the embedding $H_{\frac{1}{2}} \hookrightarrow H$ and the assumptions made in Theorem 2.5.1 hold true, the proof of the previous theorem shows thanks to [18] that the semigroup $e^{t\mathcal{A}}$ is at least dissipating over the time as $t^{-\frac{1}{1-\alpha}}$. In the following section, we will show that under some assumptions the semigroup $e^{t\mathcal{A}}$ is decreasing over the time as $t^{-\frac{1}{1-\alpha}}$. Hence, a sharp decay rate of the energy of system (2.1) holds.

2.6 Non-Uniform Stabilization

This section is devoted to study the non-uniform stabilization of system (2.1)–(2.7). Under some assumptions on the behavior of an auxiliary dissipative operator whose dissipation is generated by the classical BB^* operator, we prove a polynomial decay result for the system (2.1)–(2.7). For this purpose, we will use a frequency domain approach.

Proposition 2.6.1 *We suppose that $\eta = 0$, then the operator $-\mathcal{A}$ is not onto, and consequently $0 \in \sigma(\mathcal{A})$.*

Proof Let $Y = (0, 0, h(\xi)) \in \mathcal{H}$ and assume that there exists $X = (u, v, \varphi) \in \mathcal{D}(\mathcal{A})$ such that

$$-\mathcal{A}X = Y.$$

It follows that $v = 0$, $\varphi(\xi) = \dfrac{h(\xi)}{|\xi|^2}$ and $Au + \gamma B \displaystyle\int_{\mathbb{R}} \dfrac{p(\xi)h(\xi)}{|\xi|^2} \, d\xi = 0$. Let $\psi \in U$ such that $\psi \neq 0$ and we set $h(\xi) = \dfrac{1}{(1 + |\xi|)} \psi$. It is clear that $h \in L^2(\mathbb{R}; U)$. However, $\varphi \notin L^2(\mathbb{R}; U)$. Thus, the operator $-\mathcal{A}$ is not onto. This completes the proof. $\qquad \square$

Lemma 2.6.1 *Let $\omega \in \mathbb{R}^*$, and then for any fixed $\eta > 0$ and $0 < \alpha < 1$, we have*

$$\int_0^{+\infty} \frac{\rho^{2\alpha-1}}{(\rho^2 + \eta)^2 + \omega^2} \, d\rho = \begin{cases} \dfrac{\sin(2(\alpha - 1)(\pi - \phi)) - \sin(2(\alpha - 1)\phi)}{\sin(2\alpha\pi)\sin(2\phi)(\eta^2 + \omega^2)^{1-\frac{\alpha}{2}}} & \text{if } \alpha \neq \dfrac{1}{2} \\[3mm] \dfrac{3(2\pi - \phi)}{8(\eta^2 + \omega^2)^{\frac{3}{4}}} & \text{if } \alpha = \dfrac{1}{2}, \end{cases}$$
(2.37)

where we have denoted $\phi = \arccos\left(\dfrac{\sqrt{\dfrac{\sqrt{\eta^2+\omega^2}-\eta}{2}}}{(\eta^2 + \omega^2)^{\frac{1}{4}}} \right)$.

Proof This proof is the same as the one of Lemma 2.5.1. By keeping the same notation here, we just sketch the proof.

Case 1: $\alpha \neq \frac{1}{2}$. We set the complex function

$$f(z) = \frac{z^{2\alpha-1}}{(z^2 + \eta)^2 + \omega^2}, \qquad \forall z \in \mathbb{C} \setminus \mathbb{R}_-,$$

whose poles are $z_1^{\pm} = (\eta^2 + \omega^2)^{\frac{1}{4}}e^{\pm i\phi}$, $z_2^{\pm} = (\eta^2 + \omega^2)^{\frac{1}{4}}e^{\pm i(\phi-\pi)}$, and eventually $z_3 = 0$. Using the same arguments as Lemma 2.5.1, we can show that

$$\int_{\gamma_r} f(z)\, dz \longrightarrow 0 \quad \text{as} \quad r \searrow 0,$$

and

$$\int_{\gamma_R} f(z)\, dz \longrightarrow 0 \quad \text{as} \quad R \nearrow +\infty,$$

where on the segments $[AB]$ and $[CD]$, we have

$$\int_{\gamma_{AB}} f(z)\, dz = \int_r^R \frac{-e^{2i\alpha(\pi-\epsilon)}\rho^{2\alpha-1}}{(\rho^2 e^{2i\epsilon}+\eta)^2+\omega^2}\, d\rho \xrightarrow[\epsilon\to 0]{} \int_r^R \frac{-e^{2i\alpha\pi}\rho^{2\alpha-1}}{(\rho^2+\eta)^2+\omega^2}\, d\rho$$

and

$$\int_{\gamma_{CD}} f(z)\, dz = \int_r^R \frac{e^{2i\alpha(\epsilon-\pi)}\rho^{2\alpha-1}}{(\rho^2 e^{2i(\epsilon-\pi)}+\eta)^2+\omega^2}\, d\rho \xrightarrow[\epsilon\to 0]{} \int_r^R \frac{e^{-2i\alpha\pi}\rho^{2\alpha-1}}{(\rho^2+\eta)^2+\omega^2}\, d\rho.$$

Summing all these integrals and applying the residues theorem, we obtain

$$\int_0^{+\infty} \frac{-\sin(2\alpha\pi)\rho^{2\alpha-1}}{(\rho^2+\eta)^2+\omega^2}\, d\rho = \operatorname*{Res}_{z=z_1^{\pm},z_2^{\pm}} [f(z)]$$

$$= \frac{(\eta+\omega^2)^{\frac{\alpha}{2}}(\sin(2(\alpha-1)(\pi-\phi))-\sin(2(\alpha-1)\phi))}{2\cos(\phi)},$$

which leads obviously to the first line of (2.37).

Case 2: $\alpha = \frac{1}{2}$. In this case we have just to remark that

$$\frac{1}{(\rho^2+\eta)^2+\omega^2} = \frac{1}{8\tau^3\cos(\phi)}\left[\frac{\rho+\tau\cos(\phi)}{\rho^2+2\tau\cos(\phi)\rho+\tau^2}-[\frac{\rho-\tau\cos(\phi)}{\rho^2-2\tau\cos(\phi)\rho+\tau^2}\right]$$

$$+ 6\tau\cos(\phi)\left[\frac{1}{\rho^2+2\tau\cos(\phi)\rho+\tau^2}+\frac{1}{\rho^2-2\tau\cos(\phi)\rho+\tau^2}\right],$$

where we have denoted $\tau = (\eta^2+\omega^2)^{\frac{1}{4}}$. $\qquad\square$

Let us define now $\mathcal{H}_0 = H_{\frac{1}{2}} \times H$, and let us consider the operator

$$\mathcal{A}_0 : \mathcal{D}(\mathcal{A}_0) \subset \mathcal{H}_0 \longrightarrow \mathcal{H}_0$$

defined by

$$\mathcal{A}_0 = \begin{pmatrix} 0 & I \\ -A & -BB^* \end{pmatrix}$$

with domain

$$\mathcal{D}(\mathcal{A}_0) = \left\{ (w, v) \in \mathcal{H}_0 : v \in H_{\frac{1}{2}}, \ Aw + BB^*v \in H \right\}.$$

Proposition 2.6.2 *The operator $\mathcal{A}_0$ generates a C_0-semigroup of contractions in the Hilbert space $\mathcal{H}_0$. Moreover, the following auxiliary problem:*

$$\begin{cases} \partial_t^2 w(t) + Aw + BB^* \partial_t w(t) = 0 \\ w(0) = w^0, \ \partial_t w(0) = w^1 \end{cases} \tag{2.38}$$

admits a unique solution $w(t, x)$ in such a way that if $(w^0, w^1) \in \mathcal{D}(\mathcal{A}_0)$ the solution $w(t, x)$ of (2.38) verifying the following regularity:

$$(w, \partial_t w) \in \mathcal{C}([0, +\infty); \mathcal{D}(\mathcal{A}_0)) \cap \mathcal{C}^1([0, +\infty); \mathcal{H}_0)$$

and when $(w^0, w^1) \in \mathcal{H}_0$, we have

$$(w, \partial_t w) \in \mathcal{C}([0, +\infty); \mathcal{H}_0).$$

The energy of the system (2.38) defined as follows:

$$E_0(t) = \frac{1}{2} \left(\|\partial_t w(t)\|_H^2 + \|w(t)\|_{H_{\frac{1}{2}}}^2 \right)$$

is decreasing in time, and in particular we have

$$\frac{dE_0}{dt}(t) = -\|B^* \partial_t w(t)\|_U^2. \tag{2.39}$$

Proof To show that $\mathcal{A}_0$ generates a C_0-semigroup of contractions, we have to prove according to the Lumer–Phillips theorem (see [47, Theorem 4.3]) that $\mathcal{A}_0$ is m-dissipative. First, let $(w, v) \in \mathcal{D}(\mathcal{A}_0)$, and then we have

$$\text{Re} \left\langle \mathcal{A}_0 \begin{pmatrix} w \\ v \end{pmatrix}, \begin{pmatrix} w \\ v \end{pmatrix} \right\rangle_{\mathcal{H}_0} = -\|B^*v\|_U^2 \le 0,$$

which proves that $\mathcal{A}_0$ is dissipative. Now it remains to prove that the range of $I - \mathcal{A}_0$ is $\mathcal{H}_0$. For this purpose, we let $(f, g) \in \mathcal{H}_0$ and we look for a couple $(w, v) \in \mathcal{D}(\mathcal{A}_0)$ such that

$$(I - \mathcal{A}_0) \begin{pmatrix} w \\ v \end{pmatrix} = \begin{pmatrix} f \\ g \end{pmatrix},$$

or equivalently,

$$\begin{cases} v = w + f \\ Aw + w + BB^*w = g - f - BB^*f. \end{cases} \tag{2.40}$$

We consider now the following bilinear form on $H_{\frac{1}{2}} \times H_{\frac{1}{2}}$ defined by

$$L(w, \psi) = \langle w, \psi \rangle_{H_{\frac{1}{2}}} + \langle w, \psi \rangle_H + \langle B^*w, B^*\psi \rangle_U.$$

It is clear that L is continuous and coercive form on $H_{\frac{1}{2}} \times H_{\frac{1}{2}}$, therefore according to the Lax–Milgram theorem, there exists a unique $w \in H_{\frac{1}{2}}$ such that

$$L(w, \psi) = \langle g - f, \psi \rangle_H - \langle B^*f, B^*\psi \rangle_U, \quad \forall \psi \in H_{\frac{1}{2}}.$$

Equivalently, this can be written as follows:

$$\langle Aw + BB^*(w + f), \psi \rangle_{H_{-\frac{1}{2}} \times H_{\frac{1}{2}}} = \langle g - f - w, \psi \rangle_H, \quad \forall \psi \in H_{\frac{1}{2}}.$$

In other words, $Aw + BB^*(w + f) \in H$ and we have

$$Aw + w + BB^*w = g - f - BB^*f.$$

Since $v = w + f$, then $v \in H^{\frac{1}{2}}$. Hence, system (2.40) admits a unique solution $(w, v) \in \mathcal{D}(\mathcal{A}_0)$. Thus, the operator $\mathcal{A}_0$ is m-dissipative, and consequently the existence and the uniqueness of the solution of problem (2.38) hold with regularity as described above. Finally, a straightforward calculation gives (2.39). □

Let $M \geq 1$ be an increasing function on $\mathbb{R}_+$ and we can assume $M(s) \geq 1$ for $s \geq 1$. We assume that $i\mathbb{R} \subset \rho(\mathcal{A}_0)$ and the following growth on the resolvent

$$\limsup_{\omega \in \mathbb{R}, |\omega| \to +\infty} \|M(|\omega|)^{-1}(i\omega I - \mathcal{A}_0)^{-1}\|_{\mathcal{L}(\mathcal{H}_0)} < +\infty. \tag{2.41}$$

This means according in particular to Huang-Prüss [34, 50] and Batty and Duyck-aerts [17] (see also Borichev and Tomilov theorem [18, Theorem 2.4]), respectively, that the semigroup $e^{t\mathcal{A}_0}$ is exponentially stable if $M(|\omega|) = 1$ and polynomially stable if $M(|\omega|) = |\omega|^\ell$ for some $\ell > 0$; namely, we have

$$\|e^{t\mathcal{A}_0}\|_{\mathcal{L}(\mathcal{H}_0)} \leq Ce^{-\delta t}, \quad \forall t \geq 0,$$

for some $\delta > 0$ when $M(|\omega|) = 1$ and

$$\|e^{t\mathcal{A}_0}(w^0, w^1)\|_{\mathcal{H}_0} \leq \frac{C}{(1+t)^{\frac{1}{\ell}}} \|(w^0, w^1)\|_{\mathcal{D}(\mathcal{A}_0)}, \quad \forall t \geq 0,$$

for all $(w^0, w^1) \in \mathcal{D}(\mathcal{A}_0)$ when $M(|\omega|) = |\omega|^\ell$. However, when $M(|\omega|) = e^{K_0|\omega|}$ for some $K_0 > 0$ imply from Burq [20] that the semigroup $e^{t\mathcal{A}_0}$ is logarithmically stable, namely we have

$$\|e^{t\mathcal{A}_0}(w^0, w^1)\|_{\mathcal{H}_0} \leq \frac{C}{\log^k(2+t)}\|(w^0, w^1)\|_{\mathcal{D}(\mathcal{A}_0^k)}, \quad \forall t \geq 0,$$

for every $(w^0, w^1) \in \mathcal{D}(\mathcal{A}_0^k)$ and $k \in \mathbb{N}^*$.

Theorem 2.6.1 *We assume that* $i\mathbb{R} \subset \rho(\mathcal{A})$, *and the condition (2.41) holds. Let* $\eta > 0$, *and there exists a constant* $C > 0$ *such that*

$$\|(i\omega I - \mathcal{A})^{-1}\|_{\mathcal{L}(\mathcal{H})} \leq C|\omega|^{1-\alpha}M(|\omega|), \quad \forall \omega \geq 1.$$

Since $i\mathbb{R} \subset \rho(\mathcal{A})$, then according to the Borichev and Tomilov theorem [18, Theorem 2.4], we obtain the following corollary.

Corollary 2.6.1 *We assume that condition (2.41) holds with* $M(|\omega|) = |\omega|^\ell$ *for* $\ell \geq 0$. *Then the semigroup* $e^{t\mathcal{A}}$ *is polynomially stable; namely, there exists a constant* $C > 0$ *such that*

$$\|e^{t\mathcal{A}}(u^0, u^1, \varphi^0)\|_{\mathcal{H}} \leq \frac{C}{(1+t)^{\frac{1}{1-\alpha+\ell}}}\|(u^0, u^1, \varphi^0)\|_{\mathcal{D}(\mathcal{A})}, \quad \forall t \geq 0,$$

for every initial data $(u^0, u^1, \varphi^0) \in \mathcal{D}(\mathcal{A})$. *In particular, the energy of the strong solution of (2.1)–(2.7) satisfies the following estimate:*

$$E(t) \leq \frac{C}{(1+t)^{\frac{2}{1-\alpha+\ell}}}\|(u^0, u^1, 0)\|^2_{\mathcal{D}(\mathcal{A})}.$$

From Batty and Duyckaerts [17], see also Burq [20] for similar result, we obtain the following corollary.

Corollary 2.6.2 *We assume that the condition (2.41) holds with* $M(|\omega|) = e^{K_0|\omega|}$ *for some* $K_0 > 0$. *Then the semigroup* $e^{t\mathcal{A}}$ *is logarithmically stable, and there exists a constant* $C > 0$ *such that*

$$\|e^{t\mathcal{A}}(u^0, u^1, \varphi^0)\|_{\mathcal{H}} \leq \frac{C}{\ln(2+t)}\|(u^0, u^1, \varphi^0)\|_{\mathcal{D}(\mathcal{A})}, \quad \forall t \geq 0,$$

for every initial data $(u^0, u^1, \varphi^0) \in \mathcal{D}(\mathcal{A})$. *In particular, the energy of the strong solution of (2.1)–(2.7) satisfies the following estimate:*

$$E(t) \leq \frac{C}{\ln^2(2+t)}\|(u^0, u^1, 0)\|^2_{\mathcal{D}(\mathcal{A})}.$$

Proof We need just to prove that

$$\limsup_{\omega \in \mathbb{R}, |\omega| \to +\infty} |\omega|^{\alpha-1} M(|\omega|)^{-1} \|(i\omega I - \mathcal{A})^{-1}\|_{\mathcal{L}(\mathcal{H})} < +\infty \qquad (2.42)$$

is satisfied. For this purpose, we argue by contradiction. We suppose that (2.42) is false, and then there exist a real sequence (ω_n), with $\omega_n \longrightarrow +\infty$ and a sequence $(u_n, v_n, \varphi_n) \in \mathcal{D}(\mathcal{A})$, verifying the following condition:

$$\|(u_n, v_n, \varphi_n)\|_{\mathcal{H}} = 1 \qquad (2.43)$$

and

$$\omega_n^{1-\alpha} M(\omega_n)(i\omega_n I - \mathcal{A}) \begin{pmatrix} u_n \\ v_n \\ \varphi_n \end{pmatrix} = \begin{pmatrix} f_n \\ g_n \\ h_n \end{pmatrix} \longrightarrow 0 \text{ in } \mathcal{H}. \qquad (2.44)$$

Multiplying (2.44) by $\begin{pmatrix} u_n \\ v_n \\ \varphi_n \end{pmatrix}$ and taking the real part of the inner product, we obtain

$$\text{Re}\left\langle \begin{pmatrix} f_n \\ g_n \\ h_n \end{pmatrix}, \begin{pmatrix} u_n \\ v_n \\ \varphi_n \end{pmatrix} \right\rangle_{\mathcal{H}} = \omega_n^{1-\alpha} M(\omega_n) \gamma \int_{\mathbb{R}} (|\xi|^2 + \eta) \|\varphi_n(\xi)\|_U^2 \, d\xi \xrightarrow[n \to +\infty]{} 0.$$

$$(2.45)$$

Detailing Eq. (2.44), we get

$$\omega_n^{1-\alpha} M(\omega_n)(i\omega_n u_n - v_n) = f_n \longrightarrow 0 \text{ in } H_{\frac{1}{2}}, \qquad (2.46)$$

$$\omega_n^{1-\alpha} M(\omega_n)\left(i\omega_n v_n + A u_n + \gamma B \int_{\mathbb{R}} p(\xi)\varphi_n(\xi) \, d\xi\right) = g_n \longrightarrow 0 \text{ in } H,$$

$$(2.47)$$

$$\omega_n^{1-\alpha} M(\omega_n)(i\omega_n \varphi_n + (|\xi|^2 + \eta)\varphi_n - p(\xi)B^* v_n) = h_n \longrightarrow 0 \text{ in } V. \qquad (2.48)$$

From (2.46) and as $\omega_n^{1-\alpha} M(\omega_n) \longrightarrow \infty$, we have

$$\omega_n \|u_n\|_H = O(1). \qquad (2.49)$$

Taking the inner product of (2.47) with u_n in H and using (2.46), one has

$$\|u_n\|_{H_{\frac{1}{2}}}^2 - \omega_n^2 \|u_n\|_H^2 = -\gamma \left\langle \int_{\mathbb{R}} p(\xi)\varphi_n(\xi) \, d\xi, B^* u_n \right\rangle_U$$

$$+ \omega_n^{\alpha-1} M(\omega_n)^{-1} (\langle g_n, u_n \rangle_H + i\omega_n \langle f_n, u_n \rangle_H).$$

Using the Cauchy–Schwarz inequality, we obtain

$$\left| \|u_n\|^2_{H_{\frac{1}{2}}} - \omega^2_n \|u_n\|^2_H \right| \le \omega^{\alpha-1}_n M(\omega_n)^{-1} \|u_n\|_H (|\omega_n|.\|f_n\|_H + \|g_n\|_H)$$

$$+\gamma \|B^* u_n\|_U \left(\int_{\mathbb{R}} \frac{p(\xi)^2}{|\xi|^2 + \eta} \, d\xi \right)^{\frac{1}{2}} \left(\int_{\mathbb{R}} (|\xi|^2 + \eta) \|\varphi_n(\xi)\|^2_U \, d\xi \right)^{\frac{1}{2}}.$$

Then, (2.43)–(2.45) and (2.49) lead to

$$\|u_n\|_{H_{\frac{1}{2}}} - \omega_n \|u_n\|_H \xrightarrow[n\to+\infty]{} 0. \tag{2.50}$$

Following (2.46), Eqs. (2.47) and (2.48) can be recast as follows:

$$\varphi_n(\xi) = i\omega_n \frac{p(\xi)}{|\xi|^2 + \eta + i\omega_n} B^* u_n - \omega^{\alpha-1}_n M(\omega_n)^{-1} \frac{p(\xi)}{|\xi|^2 + \eta + i\omega_n} B^* f_n$$

$$+ \omega^{\alpha-1}_n M(\omega_n)^{-1} \frac{h_n(\xi)}{|\xi|^2 + \eta + i\omega_n} \tag{2.51}$$

and

$$- \omega^2_n u_n + A u_n + i\omega_n \gamma \int_{\mathbb{R}} \frac{p(\xi)^2}{|\xi|^2 + \eta + i\omega_n} \, d\xi \, B B^* u_n$$

$$= \omega^{\alpha-1}_n M(\omega_n)^{-1} (g_n + i\omega_n f_n) + \gamma \omega^{\alpha-1}_n M(\omega_n)^{-1} \int_{\mathbb{R}} \frac{p(\xi)^2}{|\xi|^2 + \eta + i\omega_n} \, d\xi \, B B^* f_n$$

$$- \omega^{\alpha-1}_n M(\omega_n)^{-1} \gamma \int_{\mathbb{R}} \frac{p(\xi) B h_n(\xi)}{|\xi|^2 + \eta + i\omega_n} \, d\xi.$$

Multiplying (2.51) by $p(\xi)$, then integrating on $\mathbb{R}$ with respect to the ξ variable, and applying the Cauchy–Schwarz inequality, we obtain

$$\omega_n \left| \int_{\mathbb{R}} \frac{p(\xi)^2}{|\xi|^2 + \eta + i\omega_n} \, d\xi \right| \|B^* u_n\|_U \le \omega^{\alpha-1}_n M(\omega_n)^{-1} \left| \int_{\mathbb{R}} \frac{p(\xi)^2}{|\xi|^2 + \eta + i\omega_n} \, d\xi \right| \|B^* f_n\|_U$$

$$+ \omega^{\alpha-1}_n M(\omega_n)^{-1} \left(\int_{\mathbb{R}} \frac{p(\xi)^2}{(|\xi|^2 + \eta)^2 + \omega^2_n} \, d\xi \right)^{\frac{1}{2}} \|h_n\|_V$$

$$+ \left(\int_{\mathbb{R}} \frac{p(\xi)^2}{|\xi|^2 + \eta} \, d\xi \right)^{\frac{1}{2}} \left(\int_{\mathbb{R}} (|\xi|^2 + \eta) \|\varphi_n(\xi)\|^2_U \, d\xi \right)^{\frac{1}{2}}.$$

Using Lemmas 2.5.1 and 2.6.1, it follows

$$\omega_n(\omega_n + \eta)^{\alpha-1}\|B^*u_n\|_U \le C\left(\left(\int_{\mathbb{R}} \frac{p(\xi)^2}{|\xi|^2 + \eta}\, d\xi\right)^{\frac{1}{2}} \left(\int_{\mathbb{R}} (|\xi|^2 + \eta)\|\varphi_n(\xi)\|_U^2\, d\xi\right)^{\frac{1}{2}}\right.$$

$$\left. +\omega_n^{\alpha-1} M(\omega_n)^{-1}(\omega_n + \eta)^{\frac{\alpha}{2}-1}\|h_n\|_V + \omega_n^{\alpha-1} M(\omega_n)^{-1}(\omega_n + \eta)^{\alpha-1}\|B^*f_n\|_U\right),$$

which implies from (2.45) that

$$\omega_n^2 M(\omega_n)\|B^*u_n\|_U^2 \xrightarrow[n\to+\infty]{} 0. \tag{2.52}$$

Now we recall that the semigroup generated by the operator $\mathcal{A}_0$ is stable (in the sense of condition (2.41)) in the Hilbert space $\mathcal{H}_0$, and then there exists a unique couple $(w_n, z_n) \in \mathcal{D}(\mathcal{A}_0)$ such that

$$\begin{cases} -\omega_n^2 w_n + A w_n + i\omega_n BB^* w_n = u_n \\ z_n = i\omega_n w_n \end{cases} \tag{2.53}$$

satisfying the following estimate:

$$\omega_n\|w_n\|_H + \|w_n\|_{H_{\frac{1}{2}}} \le CM(\omega_n)\|u_n\|_H, \tag{2.54}$$

since the resolvent of $\mathcal{A}_0$ satisfies condition (2.41). Next, we take the inner product in H of the first line of (2.53) with $\omega_n w_n$, and then one gets

$$-\omega_n\|\omega_n w_n\|_H^2 + \omega_n\|w_n\|_{H_{\frac{1}{2}}}^2 + i\omega_n^2\|B^*w_n\|_U^2 = \omega_n\langle u_n, w_n\rangle_H. \tag{2.55}$$

Taking the imaginary part of (2.55) and then by using the Cauchy–Schwarz inequality and (2.54), one gets

$$\omega_n^2\|B^*w_n\|_U^2 \le \omega_n\|u_n\|_H\|w_n\|_H \le CM(\omega_n)\|u_n\|_H^2. \tag{2.56}$$

Taking the inner product of (2.52) with $\omega_n^2 w_n$ in the Hilbert space H, we have from (2.53)

$$\omega_n^2\|u_n\|_H^2 = \omega_n^{\alpha+1} M(\omega_n)^{-1}\langle g_n, w_n\rangle_H$$

$$+ i\omega_n^{\alpha+2} M(\omega_n)^{-1}\langle f_n, w_n\rangle_H$$

$$- i\omega_n^3 \gamma \int_{\mathbb{R}} \frac{p(\xi)^2}{|\xi|^2 + \eta + i\omega_n}\, d\xi \langle B^*u_n, B^*w_n\rangle_U$$

$$+ i\omega_n^3\langle B^*u_n, B^*w_n\rangle_U$$

$$+ \omega_n^{\alpha+1} M(\omega_n)^{-1} \gamma \int_{\mathbb{R}} \frac{p(\xi)^2}{|\xi|^2 + \eta + i\omega_n} \, d\xi \, \langle B^* f_n, B^* w_n \rangle_U$$

$$- \omega_n^{\alpha+1} M(\omega_n)^{-1} \gamma \int_{\mathbb{R}} p(\xi) \frac{\langle h_n(\xi), B^* w_n \rangle_U}{|\xi|^2 + \eta + i\omega_n} \, d\xi. \tag{2.57}$$

Using Lemmas 2.5.1, 2.6.1, and estimates (2.43), (2.44), (2.49), (2.52), (2.54), and (2.56), we obtain

$$|\omega_n^3 \langle B^* u_n, B^* w_n \rangle_U| \leq \omega_n^3 \|B^* u_n\|_U \|B^* w_n\|_U \leq C \omega_n^2 M(\omega_n)^{1/2} \|B^* u_n\|_U \|u_n\|_H$$

$$\leq C \omega_n M(\omega_n)^{1/2} \|B^* u_n\|_U . \omega_n \|u_n\|_H \xrightarrow[n \to +\infty]{} 0, \tag{2.58}$$

$$\left| \omega_n^3 \int_{\mathbb{R}} \frac{p(\xi)^2}{|\xi|^2 + \eta + i\omega_n} \, d\xi \, \langle B^* u_n, B^* w_n \rangle_U \right|$$

$$\leq C \omega_n^{2+\alpha} \|B^* u_n\|_U \|B^* w_n\|_U$$

$$\leq C \omega_n^\alpha M(\omega_n)^{1/2} \|B^* u_n\|_U . \omega_n \|u_n\|_H \xrightarrow[n \to +\infty]{} 0, \tag{2.59}$$

$$\omega_n^{\alpha+1} M(\omega_n)^{-1} \left| \int_{\mathbb{R}} \frac{p(\xi)^2}{|\xi|^2 + \eta + i\omega_n} \, d\xi \, \langle B^* f_n, B^* w_n \rangle_U \right|$$

$$\leq C \omega_n^{2\alpha-2} M(\omega_n)^{-1/2} \|f_n\|_{H_{\frac{1}{2}}} . \omega_n \|u_n\|_H \xrightarrow[n \to +\infty]{} 0, \tag{2.60}$$

$$\omega_n^{\alpha+1} M(\omega_n)^{-1} \left| \int_{\mathbb{R}} p(\xi) \frac{\langle h_n(\xi), B^* w_n \rangle_U}{|\xi|^2 + \eta + i\omega_n} \, d\xi \right|$$

$$\leq \omega_n^{\alpha-1} M(\omega_n)^{-1} \left(\int_{\mathbb{R}} \frac{p(\xi)^2}{(|\xi|^2 + \eta)^2 + \omega_n^2} \, d\xi \right)^{\frac{1}{2}} \|h_n\|_V . \omega_n \|u_n\|_H$$

$$\leq C \omega_n^{\alpha-1} \omega_n^{\alpha/2-1} M(\omega_n)^{-1} \|h_n\|_V . \omega_n \|u_n\|_H \xrightarrow[n \to +\infty]{} 0, \tag{2.61}$$

and

$$\omega_n^{\alpha+1} M(\omega_n)^{-1} |\langle g_n, w_n \rangle_H| \leq C\omega_n^{\alpha-1} \|g_n\|_H . \omega_n \|u_n\|_H \xrightarrow[n\to+\infty]{} 0. \qquad (2.62)$$

Taking the inner product of the first equation of (2.53) with f_n, we obtain

$$-\omega_n^2 \langle w_n, f_n \rangle_H + \langle A^{\frac{1}{2}} w_n, A^{\frac{1}{2}} f_n \rangle_H + i\omega_n \langle B^* w_n, B^* f_n \rangle_U = \langle u_n, f_n \rangle_H.$$

This with (2.43), (2.44), (2.49), (2.54), and (2.56) gives

$$\omega_n^{\alpha+2} M(\omega_n)^{-1} |\langle w_n, f_n \rangle_H|$$

$$\leq \omega_n^\alpha M(\omega_n)^{-1} \left(\|w_n\|_{H_{\frac{1}{2}}} \|f_n\|_{H_{\frac{1}{2}}} + \|u_n\|_H \|f_n\|_H + \omega_n \|B^* w_n\|_U \|B^* f_n\|_U \right)$$

$$\leq C\omega_n^{\alpha-1} (1 + M(\omega_n)^{-1} + M(\omega_n)^{-\frac{1}{2}}) \|f_n\|_{H_{\frac{1}{2}}} . \omega_n \|u_n\|_H \xrightarrow[n\to+\infty]{} 0. \qquad (2.63)$$

It follows from the combination of (2.57) and (2.58)–(2.63) that

$$\|\omega_n u_n\|_H \xrightarrow[n\to+\infty]{} 0.$$

Thus, by (2.50), we have $\|u_n\|_{H_{\frac{1}{2}}} \xrightarrow[n\to+\infty]{} 0$. Together with (2.46) and (2.45) imply that $(u_n, v_n, \varphi_n) \xrightarrow[n\to+\infty]{} 0$, which contradicts (2.43). This completes the proof. $\square$

Remark 2.6.1 In the case where for all $\delta > 0$ and

$$\sup_{Re\lambda=\delta} \left\| \lambda B^* (\lambda^2 I + A)^{-1} B \right\|_{\mathcal{L}(U)} < \infty,$$

according to [5] (see also [7]), we can replace the hypothesis (2.41) by the following observability inequalities and we obtain the same results:

- For $M(\omega) = 1$, the assumption (2.41) is equivalent to the following exact observability inequality: there exists $T, C > 0$ such that

$$\int_0^T \left\| (0 \ B^*) e^{t\begin{pmatrix} 0 & I \\ -A & 0 \end{pmatrix}} \begin{pmatrix} u^0 \\ u^1 \end{pmatrix} \right\|_U^2 dt$$

$$\geq C \left\| (u^0, u^1) \right\|_{\mathcal{H}_0}^2 , \ \forall (u^0, u^1) \in H_1 \times H.$$

- For $M(\omega) = |\omega|^\ell$ with $\ell > 0$, the assumption (2.41) can be provided from the following weak observability inequality: there exists $T, C > 0$ such that

$$
\int_0^T \left\| (0 \; B^*) e^{t \begin{pmatrix} 0 & I \\ -A & 0 \end{pmatrix}} \begin{pmatrix} u^0 \\ u^1 \end{pmatrix} \right\|_U^2 dt
$$

$$
\geq C \left\| (u^0, u^1) \right\|_{H_{-\frac{\alpha+\ell}{2}} \times H_{-\frac{1-\alpha+\ell}{2}}}^2 , \quad \forall (u^0, u^1) \in H_1 \times H.
$$

Chapter 3
Applications to the Fractional-Damped Wave Equation

In this chapter we give some applications of the abstract results of Chap. 2 to the fractional-damped wave equation. Noting that the case of the wave equation with boundary fractional damping have treated in [44, 45] where it was proven the strong stability and the lack of uniform stabilization. However, the case of the plate equation or the beam equation with boundary fractional damping was treated in [1] where in addition of that, using the domain frequency method, it was shown that the energy is polynomially stable.

3.1 Internal Fractional-Damped Wave Equation

We consider a wave equation with an internal fractional damping in a bounded and connected domain Ω of $\mathbb{R}^n$ with smooth boundary $\Gamma = \partial\Omega$

$$\begin{cases} \partial_t^2 u(x, t) - \Delta u(x, t) + a(x)\partial_t^{\alpha,\eta} u(x, t) = 0 \text{ in } \Omega \times \mathbb{R}_+ \\ u(x, t) = 0 & \text{on } \Gamma \times \mathbb{R}_+ \\ u(x, 0) = u^0(x), \quad \partial_t u(x, 0) = u^1(x) & \text{in } \Omega, \end{cases} \quad (3.1)$$

where $a(x)$ is a nonnegative function in Ω with support $\omega_0 = \text{supp}(a)$ verifying that there exist a non-empty open subset $\widetilde{\omega}_0 \subset \Omega$ and a strictly positive constant a_0 such that

$$a(x) \geq a_0 \qquad \forall x \in \widetilde{\omega}_0.$$

According to Sect. 2.2, system (3.1) can be recast as follows:

$$\begin{cases} \partial_t^2 u(x,t) - \Delta u(x,t) + \gamma\sqrt{a(x)}\displaystyle\int_{\mathbb{R}} p(\xi)\varphi(x,t,\xi)\,d\xi = 0 \ (x,t) \in \Omega \times \mathbb{R}_+ \\ \partial_t \varphi(x,t,\xi) + (|\xi|^2 + \eta)\,\varphi(x,t,\xi) = p(\xi)\sqrt{a(x)}\partial_t u(x,t) \ (x,t,\xi) \in \Omega \times \mathbb{R}_+ \times \mathbb{R} \\ u(x,t) = 0 \hspace{5.5cm} (x,t) \in \Gamma \times \mathbb{R}_+ \\ u(x,0) = u^0(x), \quad \partial_t u(x,0) = u^1(x), \quad \varphi(x,0,\xi) = 0 \quad x \in \Omega,\ \xi \in \mathbb{R}, \end{cases}$$

$$\tag{3.2}$$

where following, Sect. 2.2, $p(\xi) = |\xi|^{\frac{2\alpha-1}{2}}$. The energy of the system is given by

$$E(t) = \frac{1}{2}\left(\|\partial_t u(t)\|_{L^2(\Omega)}^2 + \|\nabla u(t)\|_{L^2(\Omega)}^2 + \gamma \int_{\mathbb{R}} \|\varphi(t,\xi)\|_{L^2(\Omega)}^2\,d\xi\right).$$

The operator $A = -\Delta$ is strictly positive and self-adjoint operator in $H = L^2(\Omega)$ and with domain $\mathcal{D}(A) = H_0^1(\Omega) \cap H^2(\Omega)$. The operator $\mathcal{A}$ corresponding to the Cauchy problem of system (3.2) is given by

$$\mathcal{A}\begin{pmatrix} u \\ v \\ \varphi \end{pmatrix} = \begin{pmatrix} v \\ \Delta u - \gamma\sqrt{a}\displaystyle\int_{\mathbb{R}} p(\xi)\varphi(\xi)\,d\xi \\ -(|\xi|^2 + \eta)\varphi(\xi) + p(\xi)\sqrt{a}v \end{pmatrix}$$

with domain in the Hilbert space $\mathcal{H} = H_0^1(\Omega) \times L^2(\Omega) \times L^2(\mathbb{R}; L^2(\Omega))$ given by

$$\mathcal{D}(\mathcal{A}) = \Big\{(u,v,\varphi) \in \mathcal{H} : v \in H_0^1(\Omega),\ \Delta u - \gamma\sqrt{a}\int_{\mathbb{R}} p(\xi)\varphi(\xi)\,d\xi \in L^2(\Omega),$$

$$|\xi|\varphi \in L^2(\mathbb{R}; L^2(\Omega)),\ (|\xi|^2 + \eta)\varphi(\xi) - p(\xi)\sqrt{a}v \in L^2(\mathbb{R}; L^2(\Omega))\Big\}.$$

Since the embedding $H_0^1(\Omega) \hookrightarrow L^2(\Omega)$ is compact and the only solution of the following problem:

$$\begin{cases} \partial_t^2 u(x,t) - \Delta u(x,t) = 0 \ (x,t) \in \Omega \times \mathbb{R}_+ \\ \sqrt{a(x)}\partial_t u(x,t) = 0 \hspace{1.5cm} (x,t) \in \Omega \times \mathbb{R}_+ \\ u(x,t) = 0 \hspace{2.9cm} (x,t) \in \Gamma \times \mathbb{R}_+ \end{cases}$$

is the trivial solution, then according to Sect. 2.4 the semigroup $e^{t\mathcal{A}}$ is strongly stable. By the same method used for the proofs of Lemmas 3.2.1 and 3.2.3, we can obtain the following lemma.

Lemma 3.1.1 *Let $\eta > 0$, and for all $\omega \in \mathbb{R}$ the operator $(i\omega I - \mathcal{A})$ is bijective from $\mathcal{D}(\mathcal{A}_0)$ to $\mathcal{H}_0$.*

We assume that the semigroup of the operator $\mathcal{A}_0 : \mathcal{D}(\mathcal{A}_0) \subset \mathcal{H}_0 \longrightarrow \mathcal{H}_0$ defined by

$$\mathcal{A}_0 \begin{pmatrix} u \\ v \end{pmatrix} = \begin{pmatrix} v \\ \Delta u - av \end{pmatrix},$$

where $\mathcal{H}_0 = H_0^1(\Omega) \times L^2(\Omega)$ with domain

$$\mathcal{D}(\mathcal{A}_0) = \left\{ (u, v) \in \mathcal{H}_0 : \Delta u - av \in L^2(\Omega), \ v \in H_0^1(\Omega) \right\},$$

is uniformly stable in the energy space $\mathcal{H}_0$, which means that the energy of the following system:

$$\begin{cases} \partial_t^2 w(x, t) - \Delta w(x, t) + a(x)\partial_t w(x, t) = 0 \text{ in } \Omega \times \mathbb{R}_+ \\ w(x, t) = 0 \qquad\qquad\qquad\qquad\qquad \text{on } \Gamma \times \mathbb{R}_+ \\ w(x, 0) = w^0(x), \quad \partial_t w(x, 0) = w^1(x) \quad \text{in } \Omega \end{cases}$$

is exponentially stable. Noting that, this can be held if the so-called geometric control condition (GCC) is satisfied (see [16]).

Proposition 3.1.1 *Under the above assumption and for $\eta > 0$, the operator $\mathcal{A}$ generates a contraction semigroup satisfying*

$$\|e^{t\mathcal{A}}X\|_{\mathcal{H}} \le \frac{C}{(1+t)^{\frac{1}{1-\alpha}}} \|X\|_{\mathcal{D}(\mathcal{A})}, \quad \forall X \in \mathcal{D}(\mathcal{A}), \ t \ge 0,$$

for some constant $C > 0$. This means that the energy of system (3.1) is decreasing to zero as t goes to $+\infty$ as $t^{\frac{-2}{1-\alpha}}$.

Proof Following Lemma 3.1.1, the operator $(i\omega I - \mathcal{A})$ is bijective for every $\omega \in \mathbb{R}$, and then using the closed graph theorem, we follow that $i\mathbb{R} \subset \rho(\mathcal{A})$. The result follows now from Corollary 2.6.1. $\qquad\square$

Remark 3.1.1 In the case where $\Omega = (0, 1) \times (0, 1)$ and

$$a(x) = \begin{cases} 1, \ \forall x \in (0, \varepsilon) \times (0, 1), \\ 0, \ \text{elsewhere}, \end{cases}$$

where $\varepsilon > 0$ is a constant, we have according to [54] that the semigroup generated by the operator $\mathcal{A}_0$ decays as $t^{-\frac{3}{2}}$ (which is optimal). We obtain in this case from Corollary 2.6.1 that the polynomial decay rate for the semigroup $e^{t\mathcal{A}}$ is given by $t^{-\frac{2}{5-2\alpha}}$.

However, we obtain a logarithm decay rate of the semigroup $e^{t\mathcal{A}}$ as given in Corollary 2.6.2 without any geometrical condition since according to [37] the resolvent of the operator $\mathcal{A}_0$ satisfies the condition (2.41) with $M(|\omega|) = e^{K_0|\omega|}$ for some $K_0 > 0$.

Remark 3.1.2 For the Euler–Bernoulli plate equation

$$\begin{cases} \partial_t^2 u(x,t) + \Delta^2 u(x,t) + a(x)\partial_t^{\alpha,\eta} u(x,t) = 0 & \text{in } \Omega \times \mathbb{R}_+ \\ u(x,t) = \Delta u(x,t) = 0 & \text{on } \Gamma \times \mathbb{R}_+ \\ u(x,0) = u^0(x), \quad \partial_t u(x,0) = u^1(x) & \text{in } \Omega, \end{cases} \qquad (3.3)$$

we obtain the same results as the system (3.1).

Moreover, for the Euler–Bernoulli plate equation with structural damping

$$\begin{cases} \partial_t^2 u(x,t) + \Delta^2 u(x,t) + div\left(a(x)\nabla\partial_t^{\alpha,\eta} u(x,t)\right) = 0 & \text{in } \Omega \times \mathbb{R}_+ \\ u(x,t) = \Delta u(x,t) = 0 & \text{on } \Gamma \times \mathbb{R}_+ \\ u(x,0) = u^0(x), \quad \partial_t u(x,0) = u^1(x) & \text{in } \Omega, \end{cases} \qquad (3.4)$$

we obtain the same results as the system (3.1) (see [6] for the logarithmic stability).

3.2 Fractional-Kelvin-Voigt Damped Wave Equation

We consider the following damped wave system:

$$\begin{cases} \partial_t^2 u(x,t) - \Delta u(x,t) - div\left(a(x)\nabla\partial_t^{\alpha,\eta} u(x,t)\right) = 0 & (x,t) \in \Omega \times \mathbb{R}_+ \\ u(x,t) = 0 & (x,t) \in \Gamma \times \mathbb{R}_+ \\ u(x,0) = u^0(x), \quad \partial_t u(x,0) = u^1(x) & x \in \Omega, \end{cases}$$

where we have made the same notation as the previous section. Equivalently, we have

$$\begin{cases} \partial_t^2 u(x,t) - \Delta u(x,t) - \gamma div\left(\sqrt{a(x)}\int_{\mathbb{R}} p(\xi)\varphi(x,t,\xi)\,d\xi\right) = 0, & (x,t) \in \Omega \times \mathbb{R}_+ \\ \partial_t\varphi(x,t,\xi) + (|\xi|^2 + \eta)\,\varphi(x,t,\xi) = p(\xi)\sqrt{a(x)}\nabla\partial_t u(x,t), & (x,t,\xi) \in \Omega \times \mathbb{R}_+ \times \mathbb{R} \\ u(x,t) = 0, \ (x,t) \in \Gamma \times \mathbb{R}_+ \\ u(x,0) = u^0(x), \quad \partial_t u(x,0) = u^1(x), \quad \varphi(x,0,\xi) = 0, \ x \in \Omega, \ \xi \in \mathbb{R}. \end{cases} \qquad (3.5)$$

The energy of the system is given by

$$E(t) = \frac{1}{2}\left(\|\partial_t u(t)\|_{L^2(\Omega)}^2 + \|\nabla u(t)\|_{L^2(\Omega)}^2 + \gamma \int_{\mathbb{R}} \|\varphi(t,\xi)\|_{(L^2(\Omega))^n}^2 \, d\xi \right).$$

The operator $A = -\Delta$ is strictly positive and self-adjoint operator in $H = L^2(\Omega)$ and with domain $\mathcal{D}(A) = H_0^1(\Omega) \cap H^2(\Omega)$. The operator $\mathcal{A}$ corresponding to the Cauchy problem of system (3.5) is given by

$$A \begin{pmatrix} u \\ v \\ \varphi \end{pmatrix} = \begin{pmatrix} v \\ \Delta u + \gamma \operatorname{div}\left(\sqrt{a} \int_{\mathbb{R}} p(\xi)\varphi(\xi)\, d\xi \right) \\ -(|\xi|^2 + \eta)\varphi(\xi) + p(\xi)\sqrt{a}\nabla v \end{pmatrix}$$

with domain in the Hilbert space $\mathcal{H} = H_0^1(\Omega) \times L^2(\Omega) \times L^2(\mathbb{R}; (L^2(\Omega))^n)$ given by

$$\mathcal{D}(A) = \Big\{ (u, v, \varphi) \in \mathcal{H} : v \in H_0^1(\Omega),\ \Delta u + \gamma \operatorname{div}\left(\sqrt{a} \int_{\mathbb{R}} p(\xi)\varphi(\xi)\, d\xi \right) \in L^2(\Omega),$$

$$|\xi|\varphi \in L^2(\mathbb{R}; (L^2(\Omega))^n),\ (|\xi|^2 + \eta)\varphi(\xi) - p(\xi)\sqrt{a}\nabla v \in L^2(\mathbb{R}; (L^2(\Omega))^n) \Big\}.$$

Lemma 3.2.1 *For all $\omega \in \mathbb{R}$, the operator $(i\omega I - A)$ is injective.*

Proof Let $X = \begin{pmatrix} u \\ v \\ \varphi \end{pmatrix} \in \mathcal{D}(A)$ such that

$$AX = i\omega X. \tag{3.6}$$

Then the dissipation property of the operator A implies that

$$\operatorname{Re}\langle AX, X \rangle = -\gamma \int_{\mathbb{R}} (|\xi|^2 + \eta)\|\varphi(\xi)\|^2_{(L^2(\Omega))^n}\, d\xi = 0.$$

Then we deduce that

$$\varphi(\xi) = 0 \quad \text{in } (L^2(\Omega))^n \text{ a.e } \xi \in \mathbb{R},$$

and problem (3.6) becomes

$$\begin{cases} v = i\omega u & \text{in } \Omega \\ \omega^2 u + \Delta u = 0 & \text{in } \Omega \\ \sqrt{a(x)}\nabla u = 0 & \text{in supp}(a) \\ u = 0 & \text{on } \Gamma. \end{cases}$$

We denote $w_j = \partial_{x_j} u$ and we derive the second and third equations, and then one gets

$$\begin{cases} \omega^2 w_j + \Delta w_j = 0 & \text{in } \Omega \\ w_j = 0 & \text{in supp}(a). \end{cases}$$

By unique continuation theorem, we find that $w_j = 0$ in Ω, and therefore $u = 0$ in Ω since $u_{|\Gamma} = 0$ and consequently $X = 0$. Thus, the operator $(i\omega I - \mathcal{A})$ is injective.

$\square$

Lemma 3.2.2 *Assume that $\eta > 0$ and $\omega \in \mathbb{R}$, and then for any $f \in H^{-1}(\Omega)$, the following problem:*

$$\begin{cases} \omega^2 u + \Delta u + (\omega^2 c_1 + i\omega c_2)\mathrm{div}(a\nabla u) = f & \text{in } \Omega \\ u = 0 & \text{on } \Gamma \end{cases} \tag{3.7}$$

where

$$c_1 = \gamma \int_{\mathbb{R}} \frac{p(\xi)^2}{(|\xi|^2 + \eta)^2 + \omega^2}\, d\xi \quad \text{and} \quad c_2 = \gamma \int_{\mathbb{R}} \frac{p(\xi)^2(|\xi|^2 + \eta)}{(|\xi|^2 + \eta)^2 + \omega^2}\, d\xi$$

admits a unique solution $u \in H_0^1(\Omega)$.

Proof First we note that the coefficients c_1 and c_2 are well defined. We distinguish two cases:

Case 1: $\omega = 0$. In this case we use Lax–Milgram's theorem to prove the existence of a unique solution $u \in H_0^1(\Omega)$ of (3.7).

Case 2: $\omega \in \mathbb{R}^*$. Separating the real and the imaginary parts of u and f by writing $u = u_1 + iu_2$ and $f = f_1 + if_2$, we consider the following auxiliary mixed system, obtained by dropping term $\omega^2 c_1$ from (3.7):

$$\begin{cases} -\Delta u_1 - c_1\omega^2\mathrm{div}(a\nabla u_1) + c_2\omega\mathrm{div}(a\nabla u_2) = f_1 & \text{in } \Omega \\ -\Delta u_2 - c_1\omega^2\mathrm{div}(a\nabla u_2) - c_2\omega\mathrm{div}(a\nabla u_1) = f_2 & \text{in } \Omega \\ u_1 = u_2 = 0 & \text{on } \Gamma. \end{cases} \tag{3.8}$$

Consider the following bilinear form in $(H_0^1(\Omega) \times H_0^1(\Omega))^2$ defined by

$$L((u_1, u_2); (w_1, w_2))$$
$$= \int_\Omega \nabla u_1 \nabla w_1\, dx + \int_\Omega \nabla u_2 \nabla w_2\, dx + c_1\omega^2 \int_\Omega a\nabla u_1 \nabla w_1\, dx$$
$$+ c_1\omega^2 \int_\Omega a\nabla u_2 \nabla w_2\, dx - c_2\omega \int_\Omega a\nabla u_2 \nabla w_1\, dx + c_2\omega \int_\Omega a\nabla u_1 \nabla w_2\, dx.$$

It is clear that L is continuous and coercive in $(H_0^1(\Omega) \times H_0^1(\Omega))^2$, and then by Lax-Milgram's theorem there exists a unique couple $(u_1, u_2) \in H_0^1(\Omega) \times H_0^1(\Omega)$ such that for every $(w_1, w_2) \in H_0^1(\Omega) \times H_0^1(\Omega)$

$$L((u_1, u_2); (w_1, w_2)) = \langle f_1, w_1 \rangle_{H^{-1} \times H_0^1} + \langle f_2, w_2 \rangle_{H^{-1} \times H_0^1}.$$

This leads to the existence and the uniqueness of a solution of the problem (3.8) in $H_0^1(\Omega) \times H_0^1(\Omega)$.

This proves in particular that the operator $A_\omega = -\Delta - (\omega^2 c_1 + i\omega c_2)\mathrm{div}(a\nabla\,.\,)$ is invertible from $H_0^1(\Omega)$ into $H^{-1}(\Omega)$, and then the first line of (3.7) is equivalent to the following equation:

$$(\omega^2 A_\omega^{-1} - I)u = A_\omega^{-1} f. \qquad (3.9)$$

It follows from the compactness of the embedding $H_0^1(\Omega) \hookrightarrow H^{-1}(\Omega)$ that the inverse operator A_ω^{-1} is compact in $H^{-1}(\Omega)$. Let us consider the following problem:

$$\begin{cases} \omega^2 u + \Delta u + (\omega^2 c_1 + i\omega c_2)\mathrm{div}(a\nabla u) = 0 & \text{in } \Omega \\ u = 0 & \text{in } \Gamma, \end{cases} \qquad (3.10)$$

and multiplying the first line of (3.10) by $\bar{u}$ and integrating over Ω, one gets

$$\omega^2 \|u\|_{L^2(\Omega)}^2 - \|\nabla u\|_{L^2(\Omega)}^2 - (\omega^2 c_1 + i\omega c_2)\|\sqrt{a}\nabla u\|_{L^2(\Omega)}^2 = 0, \qquad (3.11)$$

and then by taking the imaginary part of (3.11), we obtain $\nabla u = 0$ in $\mathrm{supp}(a)$. Proceeding as the proof of the previous lemma, one gets $u = 0$ in Ω. This proves that the operator $(\omega^2 A_\omega^{-1} - I)$ is injective. Then, following Fredhom's alternative theorem [19, Theorem 6.6], Eq. (3.9) admits a unique solution, and therefore Eq. (3.7) admits a unique solution. $\square$

Lemma 3.2.3 *Let $\eta > 0$ and a smooth enough, and then for all $\omega \in \mathbb{R}$ the operator $(i\omega I - \mathcal{A})$ is surjective.*

Proof Let $Y = (f, g, h) \in \mathcal{H}$ and we look for an $X = (u, v, \varphi) \in \mathcal{D}(\mathcal{A})$ such that

$$(i\omega I - \mathcal{A})X = Y. \qquad (3.12)$$

Equivalently, we have

$$\begin{cases} v = i\omega u - f & \text{in } \Omega \\ \omega^2 u + \Delta u + (\omega^2 c_1 + i\omega c_2)\mathrm{div}(a\nabla u) = F & \text{in } \Omega \\ \varphi(\xi) = i\omega \dfrac{p(\xi)}{|\xi|^2 + \eta + i\omega}\sqrt{a}\nabla u - \dfrac{p(\xi)}{|\xi|^2 + \eta + i\omega}\sqrt{a}\nabla f + \dfrac{h(\xi)}{|\xi|^2 + \eta + i\omega} & \text{in } \Omega \\ u = 0 & \text{on } \Gamma, \end{cases} \qquad (3.13)$$

where c_1 and c_2 are defined in Lemma 3.2.2 and $F \in L^2(\Omega)$ is given by

$$F = (c_2 - i\omega c_1)\mathrm{div}(a\nabla f) - i\omega f - g - \gamma \mathrm{div}\left(\sqrt{a}\int_{\mathbb{R}} p(\xi)\dfrac{h(x,\xi)}{|\xi|^2 + \eta + i\omega}\,d\xi\right).$$

Since for a smooth enough $F \in H^{-1}(\Omega)$, then using Lemma 3.2.2, problem (3.13) has a unique solution $u \in H_0^1(\Omega)$, and therefore problem (3.12) has a unique solution $X \in \mathcal{D}(\mathcal{A})$.   □

We consider now the following auxiliary problem:

$$\begin{cases} \partial_t^2 w(x,t) - \Delta w(x,t) + \operatorname{div}(a(x)\nabla \partial_t w(x,t)) = 0 & \text{in } \Omega \times \mathbb{R}_+ \\ w(x,t) = 0 & \text{on } \Gamma \times \mathbb{R}_+ \\ w(x,0) = w^0(x), \quad \partial_t w(x,0) = w^1(x) & \text{in } \Omega. \end{cases} \tag{3.14}$$

Equation (3.14) is well-posed in the Hilbert space $\mathcal{H}_0 = H_0^1(\Omega) \times L^2(\Omega)$, and its solution is given by the semigroup generated by the operator $\mathcal{A}_0 : \mathcal{D}(\mathcal{A}_0) \subset \mathcal{H}_0 \longrightarrow \mathcal{H}_0$ defined by

$$\mathcal{A}_0 \begin{pmatrix} w \\ v \end{pmatrix} = \begin{pmatrix} v \\ \Delta w - \operatorname{div}(a\nabla v) \end{pmatrix}$$

with domain

$$\mathcal{D}(\mathcal{A}_0) = \left\{ (w,v) \in \mathcal{H}_0 : \ \Delta w - \operatorname{div}(a\nabla v) \in L^2(\Omega), \ v \in H_0^1(\Omega) \right\}.$$

We suppose now that there exist $\Omega_j \subset \Omega$ with piecewise smooth boundary and $x_0^j \in \mathbb{R}^n$, $j = 1, 2, \dots, J$ s.t. $\Omega_i \cap \Omega_j = \varnothing$ for any $1 \le i < j \le J$ and for some $\delta > 0$,

$$\Omega \cap \mathcal{N}_\delta \left[\left(\bigcup_{j=1}^J \Gamma_j \right) \cup \left(\Omega \setminus \bigcup_{j=1}^J \Omega_j \right) \right] \subset \omega_0,$$

where for

$$S \subset \mathbb{R}^n, \ \mathcal{N}_\delta(S) = \bigcup_{x \in S} \{ y \in \mathbb{R}^n; |x - y| < \delta \}$$

and

$$\Gamma_j = \left\{ x \in \partial \Omega_j; \ \left(x - x_0^j \right).v^j(x) > 0 \right\}$$

with v^j being the unitary outward-pointing normal vector to $\partial \Omega_j$. Under the above assumptions, Tebou in [56, Theorem 1.1] shows a non-uniform stabilization result in such a way that the energy of the system (3.14) decreases to zero as t^{-1} when t goes to the infinity for regular initial data. Consequently, according to Corollary 2.6.1, we have the following:

Proposition 3.2.1 *Under the above assumptions and for $\eta > 0$, the operator $\mathcal{A}$ generates a C_0 semigroup of contractions satisfying*

$$\|e^{t\mathcal{A}}X\|_{\mathcal{H}} \leq \frac{C}{(1+t)^{\frac{1}{2-\alpha}}}\|X\|_{\mathcal{D}(\mathcal{A})}, \quad \forall X \in \mathcal{D}(\mathcal{A}), \ t \geq 0,$$

for some constant $C > 0$. This means that the energy of system (3.5) is decreasing to zero as t goes to $+\infty$ as $t^{-\frac{2}{2-\alpha}}$.

If in addition to the above geometric condition, we assume the following regularity of the damping coefficient that is $a \in W^{1,\infty}(\Omega)$, $|\nabla a(x)|^2 \leq M_0 a(x)$ a.e. in Ω, $a(x) \geq a_0$ a.e. in Ω, Tebou in [56, Theorem 1.2] (see also [57, Theorem 1.2]) shows that the semigroup generated by the operator $\mathcal{A}_0$ is uniformly stable, that is, the energy of the system (3.14) is exponentially stable. Then, by combining this with Corollary 2.6.1, we get the following proposition:

Proposition 3.2.2 *Under the above additional assumptions and for $\eta > 0$, the operator $\mathcal{A}$ generates a C_0 semigroup of contractions satisfying*

$$\|e^{t\mathcal{A}}X\|_{\mathcal{H}} \leq \frac{C}{(1+t)^{\frac{1}{1-\alpha}}}\|X\|_{\mathcal{D}(\mathcal{A})}, \quad \forall X \in \mathcal{D}(\mathcal{A}), \ t \geq 0,$$

for some constant $C > 0$. This means that the energy of system (3.5) is decreasing to zero as t goes to $+\infty$ as $t^{-\frac{2}{1-\alpha}}$.

However, without any geometric conditions, namely a is equal to a constant d in ω_0 and equal to zero elsewhere, using the Carleman estimate, Ammari, Hassine, and Robbiano [4, Theorem 1.1] show that the energy of the system (3.5) decreases to zero as $\ln^{-2}(t)$ as t goes to $+\infty$ for regular initial data. In particular, it is proven that the resolvent estimate of the operator $\mathcal{A}_0$ satisfies for some constant $C > 0$ large enough,

$$\|(i\omega I - \mathcal{A}_0)^{-1}\|_{\mathcal{L}(\mathcal{H}_0)} \leq Ce^{C|\omega|}, \quad \forall |\omega| \gg 1.$$

Hence, by combining this resolvent estimate with Corollary 2.6.2, we obtain the following proposition:

Proposition 3.2.3 *Under the above assumption and for $\eta > 0$, the operator $\mathcal{A}$ generates a C_0 semigroup of contractions satisfying*

$$\|e^{t\mathcal{A}}X\|_{\mathcal{H}} \leq \frac{C}{\ln(1+t)}\|X\|_{\mathcal{D}(\mathcal{A})}, \quad \forall X \in \mathcal{D}(\mathcal{A}), \ t \geq 0,$$

for some constant $C > 0$. This means that the energy of system (3.5) is decreasing to zero as t goes to $+\infty$ as $\ln^{-2}(t)$.

3.3 Pointwise Fractional-Damped String Equation

We consider the equation of the vibration of a string of length equal to 1 with a pointwise fractional damping modeled by the following equation:

$$\begin{cases} \partial_t^2 u(x,t) - u''(x,t) + \partial_t^{\alpha,\eta} u(\zeta,t)\delta_\zeta = 0 & (x,t) \in (0,1) \times \mathbb{R}_+ \\ u(0,t) = u(1,t) = 0 & t \in \mathbb{R}_+ \\ u(x,0) = u^0(x), \quad \partial_t u(x,0) = u^1(x) & x \in (0,1), \end{cases}$$

where the prime denotes the space derivative and δ_ζ is the Dirac mass concentrated in the point ζ of $(0,1)$ (see [33, 58] for the classical derivative). Equivalently, we have

$$\begin{cases} \partial_t^2 u(x,t) - u''(x,t) + \gamma \displaystyle\int_{\mathbb{R}} p(\xi)\varphi(t,\xi)\,d\xi\,\delta_\zeta = 0 & (x,t) \in (0,1) \times \mathbb{R}_+ \\ \partial_t \varphi(t,\xi) + (|\xi|^2 + \eta)\varphi(t,\xi) = p(\xi)\partial_t u(\zeta,t) & (x,t,\xi) \in (0,1) \times \mathbb{R}_+ \times \mathbb{R} \\ u(0,t) = u(1,t) = 0 & t \in \mathbb{R}_+ \\ u(x,0) = u^0(x), \partial_t u(x,0) = u^1(x), \partial_t \varphi(0,\xi) = \varphi^0(\xi) & (x,\xi) \in (0,1) \times \mathbb{R}, \end{cases}$$
(3.15)

where following the notation given in Chap. 2, we have $U = \mathbb{C}$, $H = L^2(0,1)$, $H_{\frac{1}{2}} = H_0^1(0,1)$, $H_{-\frac{1}{2}} = H^{-1}(0,1)$, $Bz = z\delta_\zeta$ for all $z \in \mathbb{C}$ and $B^*u = u(\zeta)$ for all $u \in H_0^1(0,1)$.

We consider now the operator $\mathcal{A} : \mathcal{D}(\mathcal{A}) \subset \mathcal{H} \longrightarrow \mathcal{H}$ defined by

$$\mathcal{A}\begin{pmatrix} u \\ v \\ \varphi \end{pmatrix} = \begin{pmatrix} v \\ u'' - \gamma \displaystyle\int_{\mathbb{R}} p(\xi)\varphi(\xi)\,d\xi\,\delta_\zeta \\ -(|\xi|^2 + \eta)\varphi(\xi) + p(\xi)v(\zeta) \end{pmatrix},$$

in the Hilbert space $\mathcal{H} = H_0^1(0,1) \times L^2(0,1) \times L^2(\mathbb{C};\mathbb{R})$ with domain

$$\mathcal{D}(\mathcal{A}) = \Big\{ (u,v,\varphi) \in \mathcal{H} : v \in H_0^1(0,1), \ u'' - \gamma \int_{\mathbb{R}} p(\xi)\varphi(\xi)\,d\xi\,\delta_\zeta \in L^2(0,1),$$

$$|\xi|\varphi \in L^2(\mathbb{R};\mathbb{C}), \ -(|\xi|^2 + \eta)\varphi(\xi) + p(\xi)v(\zeta) \in L^2(\mathbb{R};\mathbb{C}) \Big\}.$$

The energy of the solution of system (3.15) is given by

$$E(t) = \frac{1}{2}\left(\|\partial_t u(t)\|_{L^2(0,1)}^2 + \|u'(t)\|_{L^2(0,1)}^2 + \gamma \int_{\mathbb{R}} |\varphi(t,\xi)|^2\,d\xi \right).$$

Lemma 3.3.1 *Assume that $\eta \geq 0$, and then $i\omega I - \mathcal{A}$ is injective for every $\omega \in \mathbb{R}$ if and only if $\zeta \notin \mathbb{Q}$.*

Proof Let $W = {}^t(u, v, \varphi) \in \mathcal{D}(\mathcal{A})$ and $\omega \in \mathbb{R}$ such that

$$\mathcal{A}W = i\omega W. \tag{3.16}$$

Equivalently, we have

$$\begin{cases} v = i\omega u & \text{in } (0, 1) \\ -u'' + \gamma \displaystyle\int_{\mathbb{R}} p(\xi)\varphi(\xi)\,d\xi - i\omega v = 0 & \text{in } (0, 1) \\ -(|\xi|^2 + \eta)\varphi(\xi) + p(\xi)v(\zeta) - i\omega\varphi(\xi) = 0 \;\forall\,\xi \in \mathbb{R} \\ u(0) = u(1) = 0. \end{cases} \tag{3.17}$$

Taking the real part of the inner product of (3.16) with W in $\mathcal{H}$ and then by using the fact that $\mathcal{A}$ is dissipative, we obtain

$$\mathrm{Re}\,\langle \mathcal{A}W, W\rangle_{\mathcal{H}} = -\gamma \int_{\mathbb{R}} (|\xi|^2 + \eta)|\varphi(\xi)|^2\,d\xi = 0,$$

and then we deduce

$$\varphi = 0 \text{ a.e. in } \mathbb{R}.$$

Therefore, (3.17) is recast as follows:

$$\begin{cases} v = i\omega u & \text{in } (0, 1) \\ \omega^2 u + u'' = 0 & \text{in } (0, 1) \\ u(0) = u(1) = u(\zeta) = 0. \end{cases} \tag{3.18}$$

If $\omega = 0$, this obviously implies that $u = v = 0$, and then $\mathcal{A}$ is injective.

Next, we suppose that $\omega \neq 0$. The Helmholtz equation $\omega^2 u + u'' = 0$ with the Dirichlet boundary conditions $u(0) = u(1) = 0$ admits a non-trivial solution if and only if $\omega = k\pi$ for every $k \in \mathbb{Z}^*$; in particular we have

$$u(x) = \beta \sin(k\pi x) \quad \forall\,\beta \in \mathbb{C}.$$

By taking into account the boundary condition $u(\zeta) = 0$, then, W is not an eigenfunction if and only if $\zeta \notin \mathbb{Q}$. This completes the proof. $\qquad\square$

Lemma 3.3.2 *Assume that $\eta > 0$, $\zeta \notin \mathbb{Q}$, and $\omega \in \mathbb{R}$. Let c_1 and c_2 be the two constants defined in Lemma 3.2.2. Then, for any $f_1 \in L^2(0, \zeta)$, $f_2 \in L^2(\zeta, 1)$, the following problem:*

$$\begin{cases} \omega^2 u_1 + u_1'' = f_1 & \text{in } (0, \zeta) \\ \omega^2 u_2 + u_2'' = f_2 & \text{in } (\zeta, 1) \\ u_1(\zeta) = u_2(\zeta) \\ u_2'(\zeta) - u_1'(\zeta) = (\omega^2 c_1 + i\omega c_2) u_1(\zeta) \\ u_1(0) = u_2(1) = 0 \end{cases} \tag{3.19}$$

admits a unique solution $(u_1, u_2) \in H_L^1(0, \zeta) \cap H^2(0, \zeta) \times H_R^1(\zeta, 1) \cap H^2(\zeta, 1)$, where

$$H_L^1(0, \zeta) = \left\{ u \in H^1(0, \zeta) : u(0) = 0 \right\} \text{ and } H_R^1(\zeta, 1) = \left\{ u \in H^1(\zeta, 1) : u(1) = 0 \right\}.$$

Proof We distinguish two cases:

Case 1: $\omega = 0$. We use Lax–Miligram's theorem to prove the existence and a uniqueness of a solution $(u_1, u_2) \in H^1(0, \zeta) \cap H^2(0, \zeta) \times H^1(\zeta, 1) \cap H^2(\zeta, 1)$ of (3.19).

Case 2: $\omega \in \mathbb{R}^*$. Separating the real and the imaginary parts of u_1, u_2, f_1, and f_2 by writing $u_1 = u_{11} + i u_{12}$, $u_2 = u_{21} + i u_{22}$, $f_1 = f_{11} + i f_{12}$, and $f_2 = f_{21} + i f_{22}$, then, (3.19) is written as follows:

$$\begin{cases} \omega^2 u_{11} + u_{11}'' = f_{11} & \text{in } (0, \zeta) \\ \omega^2 u_{12} + u_{12}'' = f_{12} & \text{in } (0, \zeta) \\ \omega^2 u_{21} + u_{21}'' = f_{21} & \text{in } (\zeta, 1) \\ \omega^2 u_{22} + u_{22}'' = f_{22} & \text{in } (\zeta, 1) \\ u_{11}(\zeta) = u_{21}(\zeta) \\ u_{12}(\zeta) = u_{22}(\zeta) \\ u_{12}'(\zeta) - u_{11}'(\zeta) = \omega^2 c_1 u_{11}(\zeta) - \omega c_2 u_{12}(\zeta) \\ u_{22}'(\zeta) - u_{21}'(\zeta) = \omega^2 c_1 u_{12}(\zeta) + \omega c_2 u_{11}(\zeta) \\ u_{11}(0) = u_{12}(0) = u_{21}(1) = u_{22}(1) = 0. \end{cases} \tag{3.20}$$

We consider the following auxiliary mixed system, obtained by dropping the term $\omega^2 u_{ij}$ from (3.20):

$$\begin{cases} u_{11}'' = f_{11} & \text{in } (0, \zeta) \\ u_{12}'' = f_{12} & \text{in } (0, \zeta) \\ u_{21}'' = f_{21} & \text{in } (\zeta, 1) \\ u_{22}'' = f_{22} & \text{in } (\zeta, 1) \\ u_{11}(\zeta) = u_{21}(\zeta) \\ u_{12}(\zeta) = u_{22}(\zeta) \\ u_{12}'(\zeta) - u_{11}'(\zeta) = \omega^2 c_1 u_{11}(\zeta) - \omega c_2 u_{12}(\zeta) \\ u_{22}'(\zeta) - u_{21}'(\zeta) = \omega^2 c_1 u_{12}(\zeta) + \omega c_2 u_{11}(\zeta) \\ u_{11}(0) = u_{12}(0) = u_{21}(1) = u_{22}(1) = 0. \end{cases} \tag{3.21}$$

We set

$$V = \{(v_1, v_2) \in H^1(0, \varsigma) \times H^1(\varsigma, 1) : v_1(0) = v_2(1) = 0, \; v_1(\varsigma) = v_2(\varsigma)\},$$

and let us consider the following continuous and coercive bilinear form in $V \times V$:

$$L\Big((u_{11}, u_{21}, u_{12}, u_{22}), (w_{11}, w_{21}, w_{12}, w_{22})\Big)$$

$$= \int_0^\varsigma u'_{11} w'_{11} \, dx + \int_0^\varsigma u'_{12} w'_{12} \, dx + \int_\varsigma^1 u'_{21} w'_{21} \, dx$$

$$+ \int_\varsigma^1 u'_{22} w'_{22} \, dx + \omega^2 c_1\Big(u_{11}(\varsigma) w_{11}(\varsigma) + u_{12}(\varsigma) w_{12}(\varsigma)\Big)$$

$$- \omega c_2\Big(u_{12}(\varsigma) w_{11}(\varsigma) - u_{11}(\varsigma) w_{12}(\varsigma)\Big).$$

Then, by Lax–Milgram's theorem, there exists $(u_{11}, u_{21}, u_{12}, u_{22}) \in V \times V$ such that for every $(w_{11}, w_{21}, w_{12}, w_{22}) \in V \times V$, we have

$$L\Big((u_{11}, u_{21}, u_{12}, u_{22}), (w_{11}, w_{21}, w_{12}, w_{22})\Big) = -\int_0^\varsigma f_{11} w_{11} \, dx - \int_0^\varsigma f_{12} w_{12} \, dx$$

$$- \int_\varsigma^1 f_{21} w_{21} \, dx - \int_\varsigma^1 f_{22} w_{22} \, dx.$$

This implies the existence of a unique solution $(u_{11}, u_{12}, u_{21}, u_{22})$ of problem (3.21) in V. This proves in particular that the operator $A_\omega = \left(-\dfrac{d^2}{dx^2}, -\dfrac{d^2}{dx^2}\right)$ is invertible from

$$W = \Big\{(u_1, u_2) \in H_L^1(0, \varsigma) \times H_R^1(\varsigma, 1) : \; u_1(\varsigma) = u_2(\varsigma),$$

$$u'_1(\varsigma) - u'_2(\varsigma) = (\omega^2 c_1 + i\omega c_2) u_1(\varsigma)\Big\}$$

into $L^2(0, \varsigma) \times L^2(\varsigma, 1)$. It follows from the compactness of the embedding $H_L^1(0, \varsigma) \times H_R^1(\varsigma, 1) \hookrightarrow L^2(0, \varsigma) \times L^2(\varsigma, 1)$ that the inverse operator A_ω^{-1} is compact in $L^2(0, \varsigma) \times L^2(\varsigma, 1)$.

Let us consider the following problem:

$$\begin{cases} \omega^2 u_1 + u''_1 = 0 & \text{in } (0, \varsigma) \\ \omega^2 u_2 + u''_2 = 0 & \text{in } (\varsigma, 1) \\ u_1(\varsigma) = u_2(\varsigma) \\ u'_2(\varsigma) - u'_1(\varsigma) = (\omega^2 c_1 + i\omega c_2) u_1(\varsigma) \\ u_1(0) = u_2(1) = 0. \end{cases} \qquad (3.22)$$

Multiplying the first line of (3.22) by $\bar{u}_1$, integrating over $(0, \zeta)$, then multiplying the second line of (3.22) by $\bar{u}_2$, integrating over $(\zeta, 1)$, and summing up the two equations. By integration by parts, we obtain

$$\omega^2 \left(\|u_1\|^2_{L^2(0,\zeta)} + \|u_2\|^2_{L^2(\zeta,1)} \right) - \left(\|u_1'\|^2_{L^2(0,\zeta)} + \|u_2'\|^2_{L^2(\zeta,1)} \right) + (\omega^2 c_1 + i\omega)|u_1(\zeta)|^2 = 0. \tag{3.23}$$

Then, by taking the imaginary part of (3.23), we obtain that $u_1(\zeta) = u_2(\zeta) = 0$. Therefore, following the arguments applied in the proof of Lemma 3.3.1, one gets that $u_1 = u_2 = 0$. Consequently, the operator $(\omega^2 A_\omega^{-1} - I)$ is injective. Hence, following Fredholm's alternative theorem [19, Theorem 6.6], Eq. (3.22) admits a unique solution. Thus, (3.19) admits a unique solution $(u_1, u_2) \in H^1(0, \zeta) \cap H^2(0, \zeta) \times H^1(\zeta, 1) \cap H^2(\zeta, 1)$.   □

Lemma 3.3.3 *Let $\zeta \notin \mathbb{Q}$. Then if $\eta > 0$, for every $\omega \in \mathbb{R}$, $i\omega I - A$ is surjective on $\mathcal{H}$, and if $\eta = 0$ for every $\omega \in \mathbb{R}^*$, $i\omega I - A$ is surjective on $\mathcal{H}$.*

Proof Let $\omega \in \mathbb{R}$ and $F = (f, g, h) \in \mathcal{H}$. We look for $U = {}^t(u, v, \varphi) \in \mathcal{D}(A)$ solution of

$$(i\omega I - A)U = F, \tag{3.24}$$

or equivalently,

$$\begin{cases} v = i\omega u - f & (0, 1) \\ -\omega^2 u - u'' + (\omega^2 c_1 + i\omega c_2) u(\zeta) \delta_\zeta = g + i\omega f + (c_2 - i\omega c_1) f(\zeta) \delta_\zeta + I \delta_\zeta & (0, 1) \\ \varphi(\xi) = i\omega \dfrac{p(\xi)}{|\xi|^2 + \eta + i\omega} u(\zeta) - \dfrac{p(\xi)}{|\xi|^2 + \eta + i\omega} f(\zeta) + \dfrac{h(\xi)}{|\xi|^2 + \eta + i\omega} & \xi \in \mathbb{R} \\ u(0) = u(1) = 0, & \end{cases} \tag{3.25}$$

where $\omega^2 c_1 + i\omega c_2 = i\omega\gamma \displaystyle\int_{\mathbb{R}} \dfrac{p(\xi)^2}{|\xi|^2 + \eta + i\omega} \, d\xi$, $c_2 - i\omega c_1 = \gamma \displaystyle\int_{\mathbb{R}} \dfrac{p(\xi)^2}{|\xi|^2 + \eta + i\omega} \, d\xi$, and $I = -\gamma \displaystyle\int_{\mathbb{R}} \dfrac{p(\xi)h(\xi)}{|\xi|^2 + \eta + i\omega} \, d\xi$. We note that it is easy to check that for $h \in L^2(\mathbb{C}, \mathbb{R})$ and $\alpha \in (0, 1)$ the integral I is well defined. From (3.25), we deduce v and φ from u. We only have to solve

$$\begin{cases} -\omega^2 u - u'' + (\omega^2 c_1 + i\omega c_2) u(\zeta) \delta_\zeta = g + i\omega f + (c_2 - i\omega c_1) f(\zeta) \delta_\zeta + I \delta_\zeta & (0, 1) \\ u(0) = u(1) = 0 & \end{cases} \tag{3.26}$$

admits a unique solution. For this aim, we set u_1, u_2, F_1 and F_2 such that $u = u_1 \mathbb{1}_{(0,\zeta)} + u_2 \mathbb{1}_{(\zeta,1)}$ and $-g - i\omega f = F_1 \mathbb{1}_{(0,\zeta)} + F_2 \mathbb{1}_{(\zeta,1)}$. Since $u \in H^1(0, 1)$, $u_1 \in H^2(0, \zeta) \cap H_L^1(0, \zeta)$, and $u_2 \in H^2(\zeta, 1) \cap H_R^1(\zeta, 1)$, then by trace formula

$u'' = u_1'' \oplus u_2'' + (u_2'(\zeta) - u_1'(\zeta))\delta_\zeta$, where $u_1'' \oplus u_2''$ is the distribution in $(0, 1)$ given by $u_1'' \mathbb{1}_{(0,\zeta)} + u_2'' \mathbb{1}_{(\zeta,1)}$, system (3.26) is recast as follows:

$$\begin{cases} \omega^2 u_1 + u_1'' = F_1 & \text{in } (0, \zeta) \\ \omega^2 u_2 + u_2'' = F_2 & \text{in } (\zeta, 1) \\ u_1(\zeta) = u_2(\zeta) \\ u_2'(\zeta) - u_1'(\zeta) = (\omega^2 c_1 + i\omega c_2)u_1(\zeta) - (c_2 - i\omega c_1)f_1(\zeta) - I \\ u_1(0) = u_2(1) = 0. \end{cases}$$

(3.27)

First, consider $(z_1, z_2) \in H_L^1(0, \zeta) \cap H^2(0, \zeta) \times H_R^1(\zeta, 1) \cap H^2(\zeta, 1)$ the unique solution of the following system:

$$\begin{cases} \omega^2 z_1 + z_1'' = 0 & \text{in } (0, \zeta) \\ \omega^2 z_2 + z_2'' = 0 & \text{in } (\zeta, 1) \\ z_1(\zeta) = z_2(\zeta) \\ z_2'(\zeta) - z_1'(\zeta) = (\omega^2 c_1 + i\omega c_2)z_1(\zeta) - (c_2 - i\omega c_1)f_1(\zeta) - I \\ z_1(0) = z_2(1) = 0. \end{cases}$$

From Lemma 3.3.2, this system admits a solution as long as $\zeta \notin \mathbb{Q}$, for all $\omega \in \mathbb{R}$. Then by setting $y_1 = u_1 - z_1$ and $y_2 = u_2 - z_2$, we get

$$\begin{cases} \omega^2 y_1 + y_1'' = F_1 & (0, \zeta) \\ \omega^2 y_2 + y_2'' = F_2 & (\zeta, 1) \\ y_1(\zeta) = y_2(\zeta) \\ y_2'(\zeta) - y_1'(\zeta) = (\omega^2 c_1 + i\omega c_2)y_1(\zeta) \\ y_1(0) = y_2(1) = 0. \end{cases}$$

(3.28)

Using Lemma 3.3.2, problem (3.28) has a unique solution $(y_1, y_2) \in H_L^1(0, \zeta) \cap H^2(0, \zeta) \times H_R^1(\zeta, 1) \cap H^2(\zeta, 1)$. Therefore, problem (3.27) has a unique solution $(u_1, u_2) \in H_L^1(0, \zeta) \cap H^2(0, \zeta) \times H_R^1(\zeta, 1) \cap H^2(\zeta, 1)$. Reporting to (3.25) we deduce that $U = {}^t(u, v, \varphi)$ belongs to $\mathcal{D}(\mathcal{A})$ and is the solution of (3.24). This completes the proof. $\square$

Proposition 3.3.1 *We assume that $\eta \geq 0$. Then the semigroup generated by the operator $\mathcal{A}$ is strongly stable, i.e.,*

$$\lim_{t \to +\infty} \|e^{\mathcal{A}t}(u^0, v^0, \varphi^0)\|_{\mathcal{H}} = 0, \qquad \forall (u^0, u^1, \varphi^0) \in \mathcal{H},$$

if and only if $\zeta \notin \mathbb{Q}$.

Proof The proof is divided into two steps:

- We suppose that $\zeta \notin \mathbb{Q}$. From Lemmas 3.3.1 and 3.3.3, we have that $\sigma(\mathcal{A}) \cap i\mathbb{R} = \varnothing$ if $\eta > 0$ and $\sigma(\mathcal{A}) \cap i\mathbb{R} \subset \{0\}$, $\sigma_p(\mathcal{A}) \cap i\mathbb{R} = \varnothing$ if $\eta = 0$, where $\sigma_p(A)$ stands for the eigenvalue of A, which implies by a general criteria of Arendt-Batty [12] that $\mathcal{A}$ is strongly stable according to which a C_0-semigroup of contractions $e^{t\mathcal{A}}$ is strongly stable, if $\sigma(A) \cap i\mathbb{R}$ is countable and no eigenvalue of $\mathcal{A}$ lies on the imaginary axis.
- We recall that the sequence of eigenfunctions of the Dirichlet Laplacian operator in $(0, 1)$ is given by

$$u_k(x) = \sin(k\pi x) \qquad \forall x \in (0, 1)$$

formed an orthonormal basis of $L^2(0, 1)$ with the corresponding eigenvalues $-\mu_k = -k^2$ for all $k \in \mathbb{N}$. If $\zeta \in \mathbb{Q}$, then $B^* u_k = \sin(k\pi \zeta) = 0$ for some $k \in \mathbb{N}$. Following the second item of Theorem 2.5.1, ik is an eigenvalue of the operator $\mathcal{A}$. Therefore $\sigma_p(\mathcal{A}) \cap i\mathbb{R} \neq \varnothing$. Thus, $\mathcal{A}$ is not strongly stable.

This completes the proof.   □

We consider now the following auxiliary problem:

$$\begin{cases} \partial_t^2 w(x, t) - w''(x, t) + \partial_t w(\zeta, t)\delta_\zeta = 0 & (x, t) \in (0, 1) \times \mathbb{R}_+ \\ w(0, t) = w(1, t) = 0 & t \in \mathbb{R}_+ \\ w(x, 0) = w^0(x), \qquad \partial_t w(x, 0) = w^1(x) \ x \in (0, 1). \end{cases} \tag{3.29}$$

System (3.29) is well-posed in the Hilbert space $\mathcal{H}_0 = H_0^1(0, 1) \times L^2(0, 1)$, and its solution is given by the semigroup generated by the operator $\mathcal{A}_0 : \mathcal{D}(\mathcal{A}_0) \subset \mathcal{H}_0 \longrightarrow \mathcal{H}_0$ defined by

$$\mathcal{A}_0 \begin{pmatrix} w \\ v \end{pmatrix} = \begin{pmatrix} v \\ w'' - v(\zeta)\delta_\zeta \end{pmatrix}$$

with domain

$\mathcal{D}(\mathcal{A}_0)$

$= \{(w, v) \in \left[H_0^1(0, 1) \right]^2 : w \in H^2(0, \zeta) \cap H^2(\zeta, 1), \ w'(\zeta^+) - w'(\zeta^-) = v(\zeta)\}.$

Consider the following subset:

$\mathcal{M} = \{\zeta \in (0, 1) : \exists K_1, K_2 > 0,$

$\times \left(\sin^2(\mu) + \sin^2(\zeta \mu). \sin^2((1 - \zeta)\mu) \right) e^{K_1 \mu} \geq K_2, \ \forall \mu \gg 1 \}.$

According to [33, Theorem 1.1], if $\zeta \in \mathcal{M}$, the energy of the system (3.29) is decreasing as $\ln^{-2}(t)$ as t goes to $+\infty$; in particular it is proven that the resolvent estimate of the operator $\mathcal{A}_0$ satisfies for some constant $C > 0$ large enough,

$$\|(i\omega I - \mathcal{A}_0)^{-1}\|_{\mathcal{L}(\mathcal{H}_0)} \leq Ce^{C|\omega|}, \quad \forall |\omega| \gg 1.$$

Therefore, by combining this estimation with Corollary 2.6.2, we obtain the following proposition.

Proposition 3.3.2 *Under the above assumption on ζ and for $\eta > 0$, the operator $\mathcal{A}$ generates a C_0 semigroup of contractions satisfying*

$$\|e^{t\mathcal{A}}X\|_{\mathcal{H}} \leq \frac{C}{\ln(2+t)}\|X\|_{\mathcal{D}(\mathcal{A})}, \quad \forall X \in \mathcal{D}(\mathcal{A}), \ t \geq 0,$$

for some constant $C > 0$. This means that the energy of system (3.5) is decreasing to zero as t goes to $+\infty$ as $\ln^{-2}(t)$.

Remarks 3.3.1 We denote by $\mathcal{S}$ the set of all numbers $\rho \in (0,1)$ such that $\rho \notin \mathbb{Q}$ and if $[0, a_1, \ldots, a_n, \ldots]$ is the expansion of ρ as a continued fraction, then (a_n) is bounded. Let us notice that $\mathcal{S}$ is obviously uncountable and, by classical results on diophantine approximation (cf. [26, p.120]), its Lebesgue measure is equal to zero. In particular, by the Euler–Lagrange theorem (cf. [36, p.57]), $\mathcal{S}$ contains all $\zeta \in (0,1)$ such that ζ is an irrational quadratic number (i.e., satisfying a second degree equation with rational coefficients).

(1) We consider the equation of the vibration of a string of length equal to 1 with a pointwise fractional damping modeled by the following equation:

$$\begin{cases} \partial_t^2 u(x,t) - u''(x,t) + \partial_t^{\alpha,\eta} u(\zeta,t)\delta_\zeta = 0 \ (x,t) \in (0,1) \times \mathbb{R}_+ \\ u(0,t) = u'(1,t) = 0 \qquad\qquad\qquad t \in \mathbb{R}_+ \\ u(x,0) = u^0(x), \qquad \partial_t u(x,0) = u^1(x) \ x \in (0,1), \end{cases}$$

and we have according to [8, 9]:

- If $\zeta \in (0,1)$ admits a coprime factorization

$$\zeta = \frac{p}{q} \text{ with } p \text{ odd,}$$

then the energy decreases to zero as $t^{-\frac{2}{1-\alpha}}$.

- For all $\zeta \in \mathcal{S}$ the energy decreases to zero as $t^{-\frac{4}{3-2\alpha}}$.
- If $\varepsilon > 0$, then, for almost all $\zeta \in (0,1)$, the energy decreases to zero as $t^{-\frac{4(1+\varepsilon)}{3-2\alpha+2(1-\alpha)\varepsilon}}$.

(2) According to [5] and for the Euler–Bernoulli beam equation with a pointwise fractional damping modeled by the following equation:

$$\begin{cases} \partial_t^2 u(x,t) + u''''(x,t) + \partial_t^{\alpha,\eta} u(\zeta,t)\delta_\zeta = 0 & (x,t) \in (0,1) \times \mathbb{R}_+ \\ u(0,t) = u(1,t) = u''(0,t) = u''(1,t) = 0 \, t \in \mathbb{R}_+ \\ u(x,0) = u^0(x), \qquad \partial_t u(x,0) = u^1(x) \quad x \in (0,1), \end{cases}$$

we have

- For all $\zeta \in S$, the energy decreases to zero as $t^{-\frac{2}{2-\alpha}}$.
- If $\varepsilon > 0$ then, for almost all $\zeta \in (0,1)$, the energy decreases to zero as $t^{-\frac{4(1+\varepsilon)}{3-2\alpha+2(1-\alpha)\varepsilon}}$.

(3) According to [11] and for the Euler–Bernoulli beam equation with a pointwise fractional damping modeled by the following equation:

$$\begin{cases} \partial_t^2 u(x,t) + u''''(x,t) + \partial_t^{\alpha,\eta} u(\zeta,t)\delta_\zeta = 0 & (x,t) \in (0,1) \times \mathbb{R}_+ \\ u(0,t) = u'(1,t) = u''(0,t) = u'''(1,t) = 0 \, t \in \mathbb{R}_+ \\ u(x,0) = u^0(x), \qquad \partial_t u(x,0) = u^1(x) \quad x \in (0,1), \end{cases}$$

we have:

- If $\zeta \in (0,1)$ admits a coprime factorization

$$\zeta = \frac{p}{q} \text{ with } p \text{ odd,}$$

then the energy decreases to zero as $t^{-\frac{2}{1-\alpha}}$.
- For all $\zeta \in S$, the energy decreases to zero as $t^{-\frac{2}{2-\alpha}}$.
- If $\varepsilon > 0$ then, for almost all $\zeta \in (0,1)$ the energy decreases to zero as $t^{-\frac{2(1+\varepsilon)}{2-\alpha+2\varepsilon}}$.

(4) According to [10] and for the Kirchhoff beam equation with force and moment pointwise fractional damping

$$\begin{cases} \partial_t^2 u(x,t) - \partial_x^2 u'' + u''''(x,t) + \partial_t^{\alpha,\eta} u(\zeta,t)\delta_\zeta \\ \quad + \partial_t^{\alpha,\eta} \partial_x u(\zeta,t)\dfrac{d\delta_\zeta}{dx} = 0, & (x,t) \in (0,1) \times \mathbb{R}_+ \\ u(0,t) = u(1,t) = u''(0,t) = u''(1,t) = 0, & t \in \mathbb{R}_+ \\ u(x,0) = u^0(x), \qquad \partial_t u(x,0) = u^1(x), & x \in (0,1), \end{cases}$$

we have:

- If $\zeta \in (0,1)$ admits a coprime factorization

$$\zeta = \frac{p}{q} \text{ with } q \text{ odd,}$$

then the energy decreases to zero as $t^{-\frac{2}{1-\alpha}}$.

- For all $\zeta \in (0, 1)$, the energy decreases to zero as $t^{-\frac{1}{1-\alpha}}$.

(5) According to [10] and for the Euler–Bernoulli beam equation with force and moment pointwise fractional damping

$$\begin{cases} \partial_t^2 u(x, t) + u''''(x, t) + \partial_t^{\alpha,\eta} u(\zeta, t)\delta_\zeta + \partial_t^{\alpha,\eta} \partial_x u(\zeta, t)\frac{d\delta_\zeta}{dx} = 0, \ (x, t) \in (0, 1) \times \mathbb{R}_+ \\ u(0, t) = u(1, t) = u''(0, t) = u''(1, t) = 0, \ t \in \mathbb{R}_+ \\ u(x, 0) = u^0(x), \ \partial_t u(x, 0) = u^1(x), \ x \in (0, 1), \end{cases}$$

we have for all $\zeta \in (0, 1)$ that the energy decreases to zero as $t^{-\frac{2}{1-\alpha}}$.

Chapter 4
Stabilization of Fractional Evolution Systems with Memory

This chapter is devoted to the analysis of some problems of stabilization of fractional (in time) partial differential equations with memory. The fractional derivative that we consider here is the Caputo derivative which depends on two parameters $\alpha \in (0, 1)$ and $\eta > 0$. Our study is concerned with two different kinds of systems with memory. More precisely, we show that the presence of the memory for the first model improves the behavior of the energy, but in the second model it seems that it is not enough to make the energy decreasing that is why we add a damping term and provide a polynomial stabilization. the solution goes to 0 when t goes to the infinity as $1/t^\alpha$.

4.1 Introduction

Let H be a Hilbert space equipped with the norm $\| \cdot \|_H$, and let $\mathcal{A} : \mathcal{D}(\mathcal{A}) \subset H \longrightarrow H$ be a closed and densely defined operator on H. We consider the following Cauchy problem described by the means of the fractional derivative and with a memory term

$$\begin{cases} \partial_t^{\alpha,\eta} u(t) = \mathcal{A}u(t) - \dfrac{\eta}{\Gamma(1-\alpha)} \displaystyle\int_0^t (t-s)^{-\alpha} e^{-\eta(t-s)} u(s) \, ds, \ t > 0, \\ u(0) = u^0, \end{cases} \tag{4.1}$$

where $\partial_t^{\alpha,\eta}$ denoted the fractional derivative defined by (2.2).

The first result of this chapter concerns the precise asymptotic behavior of the solutions of (4.1). In [28], the author considers a fractional integro-differential equation equivalent in some sense to system (4.1) but for the case $1 < \alpha < 2$ and with $\eta = 0$. He shows how the asymptotic behavior of the continuous solution depends on some parameter ω where it is assumed that $\mathcal{A}$ is a sectorial operator with a sector depending on ω. Precisely, if $\omega \geq 0$, then the continuous solutions

© The Author(s), under exclusive license to Springer Nature Switzerland AG 2022
K. Ammari et al., *Stabilization for Some Fractional-Evolution Systems*,
SpringerBriefs in Mathematics, https://doi.org/10.1007/978-3-031-17343-1_4

are bounded by an exponential of type $e^{\omega \frac{1}{\alpha} t}$, and if $\omega < 0$, then the solutions show a merely algebraic decay of order $o\left(\frac{1}{\omega t^\alpha}\right)$. In this work we prove under some consideration on the resolvent behavior (weaker than the one considered in [28]) that the second kind of behavior still true when $\eta = 0$, and one shows the effect of the memory term in the improvement of the energy decay rate. More precisely, we obtain an exponential stable decay rate when $\eta > 0$.

The second main result of this chapter is concerned with a fractional time derivative problem with memory a second kind. This problem is motivated by a temperature-dependent phase field model with memory described as follows: let Ω be an open set of $\mathbb{R}^n$, with boundary Γ of class C^4. Let ψ be the phase field (the order parameter), $\mathcal{V}$ be the temperature field, and μ be the chemical potential.

$$\begin{cases} \partial_t (\mathcal{V} + \lambda(\psi)) = a_{01} \Delta \mathcal{V} + \int_0^t a_1(t - s) \Delta \mathcal{V}(s) \, ds & \text{in } (0, +\infty) \times \Omega, \\ \mu = -\Delta \psi + \phi'(\psi) - \lambda'(\psi)\mathcal{V} & \text{in } (0, +\infty) \times \Omega, \\ \partial_t \psi = a_{02} \Delta \mu + \int_0^t a_2(t - s) \Delta \mu(s) \, ds & \text{in } (0, +\infty) \times \Omega, \\ \partial_\nu \mu = \partial_\nu \psi = \partial_\nu \mathcal{V} = 0 & \text{on } (0, +\infty) \times \Gamma, \\ \mathcal{V} = \mathcal{V}_0, \ \psi(0) = \psi_0 & \text{in } \Omega, \end{cases} \tag{4.2}$$

where a_{01} and a_{02} are nonnegative constants, $a_1, a_2 \in L^2_{\text{loc}}(\mathbb{R}^+)$ are positive and non-increasing kernels, and the physical potentials ϕ and λ are some given functions. A typical example for the kernels a_i we have in mind is given by

$$a_i(t) = \frac{t^{\alpha_i - 1}}{\Gamma(\alpha_i)} e^{-\eta_i t}, \ t > 0, \ i = 1, 2.$$

In addition, when $a_{01} = a_{02} = 0$, then equations one and three in (4.2) are of fractional time order. In [51], Prüss et al. show a convergence to steady state of solutions in the energy norm as t goes to $+\infty$ in the situation where $a_{01} = a_{02} = 0$, in particular in the time fractional case. To make such a system stable, we propose to add it a suitable damping term of the form BB^*u. Besides, we change the classical derivative into a fractional derivative. For simplicity, we choose to deal with the first equation of (4.2) only (for $\lambda = 0$) with its abstract form given as

$$\begin{cases} \partial_t^{\alpha,0} u(t) + \frac{1}{\Gamma(\alpha)} \int_0^t (t - s)^{\alpha-1} Au(s) \, ds + BB^*u(t) = 0, \ t > 0, \\ u(0) = u^0. \end{cases} \tag{4.3}$$

where A is an unbounded operator on a Hilbert space X, and B is bounded operator mapping form a Hilbert space U into $X_{-\frac{1}{2}}$. In particular, here we improve the works of Prüss et al. in [51] and Aizicovici and Petzeltová in [2] where they prove that each solution converges to a steady state as time goes to infinity for the non-isothermal Cahn–Hilliard equation with memory; in fact, here we show that (4.3) is asymptotically stable with an energy decay over the time to zero as $t^{-\alpha}$ as time goes to the infinity.

This chapter is organized as follows: in Sect. 4.2, we prove the well-posedness of system (4.1) and give an exponential stability result in case $\eta > 0$. In Sect. 4.3, we prove that the energy of system (4.1) is polynomially stable when $\eta = 0$. In Sect. 4.4, we prove the well-posedness of system (4.3) and we show a polynomial decay rate of its energy.

4.2   Well-Posedness and Exponential Stabilization

We define the convolution product of a and u by

$$a*u(t) = \int_0^t a(t-s)u(s)\,ds, \quad \forall t \in \mathbb{R}_+,\ a \in L^1_{\text{loc}}(\mathbb{R}_+),\ u \in L^p_{\text{loc}}(\mathbb{R}_+, H),\ p \in [1, +\infty[,$$

and for $\beta > 0$ the functions g_β are given by

$$g_\beta(t) = \begin{cases} \dfrac{1}{\Gamma(\beta)} t^{\beta-1} & t > 0, \\ 0 & t \leq 0. \end{cases}$$

Noting that these functions satisfy the semigroup property, namely,

$$g_\beta * g_\gamma(t) = g_{\beta+\gamma}(t), \quad \forall t > 0,\ \beta, \gamma > 0. \tag{4.4}$$

The Riemann–Liouville fractional integral of order $\alpha > 0$ is defined as follows:

$$J^\alpha u(t) = g_\alpha * u(t), \quad \forall u \in L^1_{\text{loc}}(\mathbb{R}_+; H),\ t > 0.$$

From (4.4), it follows that J^α verifying the semigroup property,

$$J^\beta J^\gamma u(t) = J^{\beta+\gamma} u(t), \quad \forall t > 0,\ \beta, \gamma > 0. \tag{4.5}$$

For every $u \in L^1_{\text{loc}}(\mathbb{R}_+; H)$ such that $g_{1-\alpha} * u \in W^{1,1}_{\text{loc}}(\mathbb{R}_+; H)$, the Riemann–Liouville fractional derivative of order α is defined by

$$D_t^\alpha u(t) = (g_{1-\alpha} * u)'(t) = (J^{1-\alpha} u)'(t), \quad \forall t > 0.$$

The operator D_t^α is the left and right inverse of J^α (see [13, Theorem 1.5]) more precisely, and we have

$$D_t^\alpha J^\alpha u(t) = u(t), \quad \forall u \in L_{loc}^1(\mathbb{R}_+; H), \ t > 0, \tag{4.6}$$

and for every $u \in L_{loc}^1(\mathbb{R}_+; H)$ such that $g_{1-\alpha} * u \in W_{loc}^{1,1}(\mathbb{R}_+; H)$,

$$J^\alpha D_t^\alpha u(t) = u(t) \quad \forall t > 0. \tag{4.7}$$

We recall that the Caputo fractional derivative of order $\alpha > 0$ when $\eta = 0$ is defined by

$$\mathbf{D}_t^\alpha u(t) = J^{1-\alpha} u'(t), \quad \forall u \in W_{loc}^{1,1}(\mathbb{R}_+; H), \ t > 0, \tag{4.8}$$

and then by integration by parts it follows that when $u \in W_{loc}^{1,1}(\mathbb{R}_+; H)$, we have

$$\mathbf{D}_t^\alpha u(t) = D_t^\alpha(u - u(0))(t), \quad \forall t > 0.$$

Hence, the Caputo derivative $\mathbf{D}_t^\alpha$ is a left inverse of J^α, but in general it is not a right inverse; namely, using (4.6) and (4.7), we have

$$\mathbf{D}_t^\alpha J^\alpha u(t) = u(t), \quad \forall u \in L_{loc}^1(\mathbb{R}_+), \ t > 0, \tag{4.9}$$

and

$$J^\alpha \mathbf{D}_t^\alpha u(t) = u(t) - u(0), \quad \forall u \in C(\mathbb{R}_+; H), \ g_{1-\alpha} * (u - u(0)) \in W_{loc}^{1,1}(\mathbb{R}_+; H), \ t > 0. \tag{4.10}$$

By setting $v(t) = u(t)e^{\eta t}$, (4.1) is equivalent to the following problem:

$$\begin{cases} \mathbf{D}_t^\alpha v(t) = \mathcal{A}v(t), \ t > 0, \\ v(0) = u^0. \end{cases} \tag{4.11}$$

Applying J_α in both sides of the first line of (4.11) and using (4.9) and (4.10), we conclude that when $u \in C(\mathbb{R}_+; H)$ satisfying $g_{1-\alpha} * (v - u^0) \in W_{loc}^{1,1}(\mathbb{R}_+; H)$, (4.11) is equivalent to the following integral differential equation:

$$u(t) = e^{-\eta t} u^0 + e^{-\eta t}(g_\alpha * (e^{\eta \cdot} \mathcal{A}u))(t), \quad \forall t \geq 0. \tag{4.12}$$

The well-posedness of an equation such as (4.1) is related to the notion of what is called solution operator defined as follows.

Definition 4.2.1 A family $(S_{\alpha,\eta}(t))_{t \geq 0} \in \mathcal{L}(H)$ (denoted simply by $(S_\alpha(t))_{t \geq 0}$ if $\eta = 0$) is called a solution operator (or a resolvent) for (4.1) or for (4.12) if the following conditions are satisfied:

(a) $S_{\alpha,\eta}(t)$ is strongly continuous for $t \geq 0$ and $S_{\alpha,\eta}(0) = I$.
(b) $S_{\alpha,\eta}(t)(\mathcal{D}(\mathcal{A})) \subset \mathcal{D}(\mathcal{A})$ and $\mathcal{A}S_{\alpha,\eta}(t)x = S_{\alpha,\eta}(t)\mathcal{A}x$ for all $x \in \mathcal{D}(\mathcal{A})$ and $t \geq 0$.

(c) The resolvent equation holds

$$S_{\alpha,\eta}(t)x = e^{-\eta t}x + e^{-\eta t}(g_\alpha * (e^{\eta \cdot}AS_{\alpha,\eta}(.)x))(t), \quad \forall x \in \mathcal{D}(A), \ t \geq 0.$$

Definition 4.2.2 A function $u \in \mathcal{C}(\mathbb{R}_+, H)$ is called a strong solution of (4.12) if $u \in \mathcal{C}(\mathbb{R}_+, \mathcal{D}(A))$ and (4.12) holds on $\mathbb{R}_+$.

Definition 4.2.3 The problem (4.12) is called well-posed if for any $u^0 \in \mathcal{D}(A)$, there is a unique strong solution $u(t, u^0)$ of (4.12), and when $u_n^0 \in \mathcal{D}(A)$ such that $u_n^0 \longrightarrow 0$ as $n \nearrow +\infty$, then $u(t, u_n^0) \longrightarrow 0$ as $n \nearrow +\infty$ in H, uniformly on compact intervals.

Definition 4.2.4 A function $u \in \mathcal{C}(\mathbb{R}_+, H)$ is called a strong solution of (4.1) if $u \in \mathcal{C}(\mathbb{R}_+, \mathcal{D}(A))$, $g_{1-\alpha} * ((e^{\eta \cdot}u) - u^0) \in \mathcal{C}^1(\mathbb{R}_+, H)$, and (4.1) holds on $\mathbb{R}_+$.

Definition 4.2.5 The problem (4.1) is called well-posed if for any $u^0 \in \mathcal{D}(A)$, there is a unique strong solution $u(t, u^0)$ of (4.1), and when $u_n^0 \in \mathcal{D}(A)$ such that $u_n^0 \longrightarrow 0$ as $n \nearrow +\infty$, then $u(t, u_n^0) \longrightarrow 0$ as $n \nearrow +\infty$ in H, uniformly on compact intervals.

Proposition 4.2.1

1. *Equation (4.12) is well-posed if and only if it admits a resolvent $S_{\alpha,\eta}(t)$. In this case we have in addition*

$$(g_\alpha * (e^{\eta \cdot}S_{\alpha,\eta}(.)x))(t) \in \mathcal{D}(A) \quad \forall x \in H \ \forall t \geq 0,$$

and we have

$$S_{\alpha,\eta}(t)x = e^{-\eta t}x + e^{-\eta t}A(g_\alpha * (e^{\eta \cdot}S_{\alpha,\eta}(.)x))(t), \quad \forall x \in H, \ t \geq 0. \quad (4.13)$$

2. *System (4.1) is well-posed if and only if (4.12) is well-posed.*

Proof

1. Problem (4.12) can be written as follows:

$$v(t) = u^0 + g_\alpha * Av(t), \quad t \geq 0, \quad (4.14)$$

where we denoted $v(t) = e^{\eta t}u(t)$. Following [49, Proposition 1.1], (4.14) is well-posed if and only if it admits a solution operator $\tilde{S}_\alpha(t)$. The result follows easily by observing that $S_{\alpha,\eta}(t) = \tilde{S}_\alpha(t)e^{-\eta t}$ is a solution operator of (4.12), and then (4.13) holds.
2. The first implication follows easily from the definitions. Now suppose that (4.12) is well-posed, and let u be its solution and let v as above solution of (4.14). To

prove the result, we only have to prove that $g_{1-\alpha} * (v - u^0) \in C^1(\mathbb{R}_+, H)$. Since $v \in C(\mathbb{R}_+, \mathcal{D}(\mathcal{A}))$, then the convolution of (4.14) with $g_{1-\alpha}$ gives

$$g_{1-\alpha} * (v - u^0)(t) = \mathcal{A} \int_0^t v(s) \, ds, \quad t \geq 0,$$

where we have used (4.4). Then it is easy to show that $g_{1-\alpha} * (v - u^0) \in C^1(\mathbb{R}_+, H)$.

This completes the proof. $\square$

Definition 4.2.6 The solution operator $S_\alpha(t)$ ($\eta = 0$) is called exponentially bounded if there are $M \geq 1$ and $\omega \geq 0$ such that

$$\|S_\alpha(t)\|_{\mathcal{L}(H)} \leq M e^{\omega t}. \tag{4.15}$$

The operator $\mathcal{A}$ is said to belong to $\mathscr{C}^\alpha(M, \omega)$ if the problem (4.1) has a solution operator $S_\alpha(t)$ satisfying (4.15). Denote $\mathscr{C}^\alpha(\omega) = \cup\{\mathscr{C}^\alpha(M, \omega) : M \geq 1\}$ and $\mathscr{C}^\alpha = \cup\{\mathscr{C}^\alpha(\omega) : \omega \geq 0\}$.

For $\theta \in [0, \pi)$, we denote

$$\Sigma_\theta = \left\{z \in \mathbb{C}^* : |\arg(z)| < \theta\right\}.$$

Definition 4.2.7 A solution operator $S_\alpha(t)$ of (4.11) is called analytic if $S_\alpha(t)$ admits an analytic extension to a sector Σ_{θ_0} for some $\theta_0 \in (0, \frac{\pi}{2}]$. An analytic solution operator is said to be of analytic type (θ_0, ω_0) if for each $\theta < \theta_0$ and $\omega > \omega_0$ there is $M = M(\theta, \omega)$ such that

$$\|S_\alpha(t)\|_{\mathcal{L}(H)} \leq M e^{\omega \mathrm{Re}(t)}, \ \forall t \in \Sigma_\theta.$$

where $\mathrm{Re}(\lambda)$ stands for the real part of λ. The set of all operators $\mathcal{A} \in \mathscr{C}^\alpha$, generating analytic solution operator $S_{\alpha,\eta}(t)$ of type (θ_0, ω_0), is denoted by $\mathscr{A}^\alpha(\theta_0, \omega_0)$. In addition, denote

$$\mathscr{A}^\alpha(\theta_0) = \bigcup\left\{\mathscr{A}^\alpha(\theta_0, \omega_0) : \omega_0 \in \mathbb{R}_+\right\} \text{ and } \mathscr{A}^\alpha = \bigcup\left\{\mathscr{A}^\alpha(\theta_0) : \theta_0 \in]0, \frac{\pi}{2}]\right\}.$$

Proposition 4.2.2 *[13, Corollary 2.17] Suppose that $\{\lambda : \mathrm{Re}(\lambda) > 0\} \subset \rho(\mathcal{A})$, and for some $C > 0$, we have*

$$\|(\lambda I - \mathcal{A})^{-1}\|_{\mathcal{L}(H)} \leq \frac{C}{\mathrm{Re}(\lambda)}, \quad \forall \lambda \in \rho(\mathcal{A}), \ \mathrm{Re}(\lambda) > 0. \tag{4.16}$$

Then, for any $\alpha \in (0, 1)$, $\mathcal{A} \in \mathscr{A}^{\alpha}\left(\min\left\{\left(\dfrac{1}{\alpha} - 1\right), 1\right\}\dfrac{\pi}{2}, 0\right).$

Theorem 4.2.1 *Suppose that $\mathcal{A}$ is an m-dissipative operator on the Hilbert space H, then $\mathcal{A}$ generates a solution operator $S_{\alpha,\eta}(t)$ and system (4.1) is well-posed. In particular, when $\eta > 0$ and $\alpha \in (0, 1)$, system (4.1) is exponentially stable, and for some $M > 0$ we have*

$$\|S_{\alpha,\eta}(t)u^0\|_H \leq Me^{-\eta t}\|u^0\|_H, \ \forall u^0 \in \mathcal{D}(\mathcal{A}), \ t \geq 0. \tag{4.17}$$

Proof Since the operator $\mathcal{A}$ is m-dissipative, then by Tucsnak and Weiss [59, Proposition 3.1.9] property (4.16) holds. According to Proposition 4.2.2, $\mathcal{A}$ is a generator of solution operator $S_{\alpha}(t)$ of (4.11), therefore Proposition 4.2.1 leads to the well-posedness of the problem (4.11), and consequently the well-posedness of (4.1) since $v(t) = u(t)e^{\eta t}$. Besides, for some constant $M > 0$, we have $v(t) = S_{\alpha}(t)u^0$ satisfy

$$\|S_{\alpha}(t)u^0\|_H \leq M\|u^0\|_H, \ \forall t \geq 0.$$

Therefore, since $u(t) = v(t)e^{-\eta t}$, the solution operator is exponential stable and so (4.17) holds. This completes the proof. $\qquad \square$

Example 4.2.1 As examples we consider the following systems:

$$\begin{cases} \partial_t^{\alpha,\eta} u(t, x) + (i\Delta + a(x))u(t, x) \\ +\dfrac{\eta}{\Gamma(1-\alpha)} \displaystyle\int_0^t (t-s)^{-\alpha}e^{-\eta(t-s)}u(s, x)\,ds = 0, & \text{for } t > 0, x \in \Omega, \\ u(t, x) = 0, & \text{for } (t, x) \in (0, +\infty) \times \partial\Omega, \\ u(0, x) = u^0(x), & \text{for } x \in \Omega, \end{cases} \tag{4.18}$$

and

$$\begin{cases} \partial_t^{\alpha,\eta} u(t, x) + (i\Delta + i)u(t, x) \\ +\dfrac{\eta}{\Gamma(1-\alpha)} \displaystyle\int_0^t (t-s)^{-\alpha}e^{-\eta(t-s)}u(s, x)\,ds = 0, & \text{for } t > 0, x \in \Omega, \\ \partial_v u(t, x) = ib(x)\,u, & \text{for } (t, x) \in (0, +\infty) \times \partial\Omega, \\ u(0, x) = u^0(x), & \text{for } x \in \Omega, \end{cases} \tag{4.19}$$

where Ω is a smooth bounded open domain of $\mathbb{R}^n$, $\partial_v = v.\nabla$ is the derivative along v, the unit normal vector pointing outward of Ω, and $a \in L^\infty(\Omega), b \in L^\infty(\partial\Omega)$ are non-identically zero and nonnegative functions.

By a direct implication of Theorem 4.2.1, we obtain for $\eta > 0$ exponential stability results for (4.18) and (4.19) without any geometric conditions (see [16] for example) on the supports of a and b.

4.3   Polynomial Stabilization

The aim of this section is to establish a polynomial stabilization result of the system (4.1) only for the case $\eta = 0$. For this purpose, we introduce first some properties of the Mittag–Leffler function [29, Chapter XVIII] and [48, chapter 1] $E_{\alpha,\beta}$ defined by

$$E_{\alpha,\beta}(z) = \sum_{n=0}^{+\infty} \frac{z^n}{\Gamma(\alpha n + \beta)} = \frac{1}{2i\pi} \int_C \frac{\mu^{\alpha-\beta} e^\mu}{\mu^\alpha - z} \, d\mu, \quad \forall z \in \mathbb{C}, \ \alpha, \beta > 0,$$

where C is a contour which starts and ends at $-\infty$ and encircles the disc

$$D = \{\mu \in \mathbb{C} : |\mu| \le |z|^{\frac{1}{\alpha}}\}$$

counter-clockwise. For short, we denote $E_\alpha(z) = E_{\alpha,1}(z)$. The first property claims (see [48, Theorem 1.6]) that for every $\beta > 0$ and $0 < \alpha < 2$, there exists a constant $c > 0$, such that

$$|E_{\alpha,\beta}(-t)| \le \frac{c}{1+t}, \quad \forall t > 0. \tag{4.20}$$

Consider also the function of Wright type ϕ_γ (see [30, 41, 62]) given by

$$\Phi_\gamma(z) = \sum_{n=0}^{+\infty} \frac{(-z)^n}{n!\Gamma(1 - \gamma(n+1))} = \frac{1}{2i\pi} \int_{C'} \mu^{\gamma-1} e^{\mu - z\mu^\gamma} \, d\mu, \quad 0 < \gamma < 1,$$

where C' is a contour which starts and ends at $-\infty$ and encircles the origin once counter-clockwise. The relationship between the Mittag–Leffler function E_γ and the function of Wright type Φ_γ is given by

$$E_\gamma(z) = \int_0^{+\infty} \Phi_\gamma(t) \, e^{zt} \, dt, \quad \forall z \in \mathbb{C}, \ 0 < \gamma < 1. \tag{4.21}$$

That is, $E_\gamma(-z)$ is the Laplace transform of Φ_γ in the whole complex plane. Therefore, Φ_γ is a probability density function,

$$\Phi_\gamma(t) \ge 0, \ \forall t > 0; \quad \text{and} \quad \int_0^{+\infty} \Phi_\gamma(t) \, dt = 1. \tag{4.22}$$

One of the main ingredients of this section is the following proposition.

Proposition 4.3.1 *[13, Theorem 3.1] Let $0 < \alpha < \beta \le 2$, $\gamma = \dfrac{\alpha}{\beta}$, and $\omega \ge 0$. If $A \in \mathscr{C}^\beta(\omega)$ then $A \in \mathscr{C}^\alpha(\omega^{\frac{1}{\gamma}})$, and the following representation holds:*

$$S_\alpha(t) = \int_0^{+\infty} \varphi_{t,\gamma}(s) S_\beta(s)\, ds, \quad \forall t > 0, \tag{4.23}$$

where $\varphi_{t,\gamma}(s) = t^{-\gamma} \Phi_\gamma(st^{-\gamma})$. The identity (4.23) holds in the strong sense.

The main result of this section is given by the following theorem.

Theorem 4.3.1 *We suppose that $\mathcal{A}$ generates a C_0-semigroup on the Hilbert space H such that the following properties hold:*

$$\operatorname{Re}(\lambda) < 0, \ \forall \lambda \in \sigma(\mathcal{A}), \quad and \quad \sup_{\operatorname{Re}(\lambda) \geq 0} \|(\lambda I - \mathcal{A})^{-1}\|_{\mathcal{L}(H)} < +\infty. \tag{4.24}$$

Then (4.1) admits a solution operator $S_\alpha(t)$ such that there exists $c > 0$

$$\|S_\alpha(t)\|_{\mathcal{L}(H)} \leq \frac{c}{1 + t^\alpha}, \quad \forall t \geq 0. \tag{4.25}$$

Proof With the assumption made on the theorem, we can apply Proposition 4.3.1 with $\beta = 1$, and of course $\gamma = \alpha$ then (4.1) admits a solution operator $S_\alpha(t)$ given by the following formula:

$$S_\alpha(t) = \int_0^{+\infty} \varphi_{t,\alpha}(s) S(s)\, ds, \quad \forall t > 0, \tag{4.26}$$

where we denoted $S_1(t)$ simply by $S(t)$ which is the C_0-semigroup generated by $\mathcal{A}$. Thanks to the assumptions (4.24), then according to [34, 50], the uniform stabilization holds, that is, there exist $\omega_0 > 0$ and $K > 0$ such that

$$\|S(t)\|_{\mathcal{L}(H)} \leq K e^{-\omega_0 t} \quad \forall t > 0. \tag{4.27}$$

Performing a change of variable in (4.26) and using (4.22) and (4.27), we obtain

$$\|S_\alpha(t)\|_{\mathcal{L}(H)} = \left\| \int_0^{+\infty} \Phi_\alpha(s) S(st^\alpha)\, ds \right\|_{\mathcal{L}(H)} \leq \int_0^{+\infty} \Phi_\alpha(s) e^{-\omega_0 st^\alpha}\, ds \quad \forall t > 0. \tag{4.28}$$

Following (4.21) and (4.28), we find

$$\|S_\alpha(t)\|_{\mathcal{L}(H)} \leq E_\alpha(-\omega_0 t^\alpha) \quad \forall t > 0.$$

Estimate (4.25) follows now from (4.20), and this completes the proof. $\qquad\square$

Example 4.3.1 As examples we consider here the same systems as above but with $\eta = 0$:

$$\begin{cases} \partial_t^\alpha u(t,x) + (i\Delta + a(x))u(t,x) = 0, \ t > 0, x \in \Omega, \\ u = 0, \ (0,+\infty) \times \partial\Omega, \\ u(0,x) = u^0(x), \ x \in \Omega, \end{cases} \qquad (4.29)$$

and

$$\begin{cases} \partial_t^\alpha u(t,x) + (i\Delta + i)u(t,x) = 0, \ t > 0, x \in \Omega, \\ \partial_\nu u = ib(x)\,u, \ (0,+\infty) \times \partial\Omega, \\ u(0,x) = u^0(x), \ x \in \Omega, \end{cases} \qquad (4.30)$$

where Ω is a smooth bounded open domain of $\mathbb{R}^n$, $\partial_\nu = \nu.\nabla$ is the derivative along ν, the unit normal vector pointing outward of Ω and $a \in L^\infty(\Omega), b \in L^\infty(\partial\Omega)$ are non-identically zero and nonnegative functions.

By a direct application of Theorem 4.3.1, we obtain polynomial stability results for (4.29) and (4.30) under geometric conditions G.C.C. (see, respectively, [37] and [16], for example) on the supports of a and b.

4.4   Extension to Some Integro-Differential Equation

Let X be a Hilbert space equipped with the norm $\| . \|_X$, and let $A : \mathcal{D}(A) \subset X \to X$ be a closed, self-adjoint, and strictly positive operator on X with dense domain. We introduce the scale of Hilbert spaces X_β, $\beta \in \mathbb{R}$, as follows: for every $\beta \geq 0$, $X_\beta = \mathcal{D}(A^\beta)$, with the norm $\|z\|_\beta = \|A^\beta z\|_X$. The space $X_{-\beta}$ is defined by duality with respect to the pivot space X as follows: $X_{-\beta} = X_\beta^*$ for $\beta > 0$. The operator A can be extended (or restricted) to each X_β, such that it becomes a bounded operator

$$A : X_\beta \to X_{\beta-1}, \quad \forall \beta \in \mathbb{R}.$$

Let a bounded linear operator $B : U \to X_{-\frac{1}{2}}$, where U is another Hilbert space which will be identified with its dual.

With definition of the Caputo derivative (4.8), we consider the following integro-differential equation:

$$\begin{cases} \mathbf{D}_t^\alpha u(t) + g_\alpha * Au(t) + BB^*u(t) = 0, \ t > 0, \\ u(0) = u^0. \end{cases} \qquad (4.31)$$

We set the Hilbert space $\mathcal{H} = X_{\frac{1}{2}} \times X$ and we consider the unbounded operator $\mathcal{A} : \mathcal{D}(\mathcal{A}) \longrightarrow \mathcal{H}$ defined by

$$\mathcal{A} = \begin{pmatrix} 0 & I \\ -A & -BB^* \end{pmatrix},$$

where $\mathcal{D}(\mathcal{A}) = \{(v, u) \in \mathcal{H} : u \in X_{\frac{1}{2}}, \ Av + BB^*u \in X\}$. It is well known (see [3, 7]) that $\mathcal{A}$ is a generator of a C_0-semigroup of contractions on $\mathcal{H}$.

Definition 4.4.1 A function $u \in C(\mathbb{R}_+, X)$ such that $g_\alpha * u \in C(\mathbb{R}_+, X_{\frac{1}{2}})$ is called a strong solution of (4.31) if the couple $\begin{pmatrix} g_\alpha * u \\ u \end{pmatrix} \in C(\mathbb{R}_+, \mathcal{D}(\mathcal{A}))$, $\begin{pmatrix} g_1 * u \\ g_{1-\alpha} * (u - u^0) \end{pmatrix} \in C^1(\mathbb{R}_+, \mathcal{H})$, and (4.31) holds on $\mathbb{R}_+$ with $u^0 \in X$.

Definition 4.4.2 The problem (4.31) is called well-posed if for any $u^0 \in X_{\frac{1}{2}}$ such that $BB^*u^0 \in X_{\frac{1}{2}}$, there is a unique strong solution $u(t, u^0)$ of (4.31), and when $u_n^0 \in \mathcal{D}(\mathcal{A})$ such that $u_n^0 \longrightarrow 0$ as $n \nearrow +\infty$, then $u(t, u_n^0) \longrightarrow 0$ in X and $g_\alpha * u(., u_n^0)(t) \longrightarrow 0$ in $X_{\frac{1}{2}}$ as $n \nearrow +\infty$ in H, uniformly on compact intervals.

Theorem 4.4.1 *Under the above assumptions made on the operator A, system (4.31) is well-posed in such a way if $u^0 \in X_{\frac{1}{2}}$ such that $BB^*u^0 \in X$, and we have the following regularity of the solution:*

$$u \in C(\mathbb{R}_+, X_{\frac{1}{2}}), \quad g_\alpha * u \in C(\mathbb{R}_+, X_{\frac{1}{2}}), \quad g_{1-\alpha} * (u - u^0) \in C^1(\mathbb{R}_+, X).$$

If, in addition, the following properties hold:

$$i\mathbb{R} \subset \rho(\mathcal{A}), \quad and \quad \limsup_{\mu \in \mathbb{R}, |\mu| \to +\infty} \|(i\mu I - \mathcal{A})^{-1}\|_{\mathcal{L}(H)} < +\infty. \tag{4.32}$$

Then for some constant $C > 0$ and for any data $u^0 \in X_{\frac{1}{2}}$, the solution $u(t)$ of (4.31) satisfies the following asymptotic estimates:

$$\|u(t)\|_X \le \frac{C}{1 + t^\alpha} \|u^0\|_X, \quad \forall t \ge 0, \tag{4.33}$$

and

$$\|g_\alpha * u(t)\|_{X_{\frac{1}{2}}} \le \frac{C}{1 + t^\alpha} \|u^0\|_X, \quad \forall t \ge 0. \tag{4.34}$$

Remark 4.4.1 We note that in the case where $B \in \mathcal{L}(U, X)$ and if $u^0 \in X_{\frac{1}{2}}$, we have immediately that $BB^*u^0 \in X$.

Proof Let us consider the following equation:

$$\begin{cases} \mathbf{D}_t^\alpha U(t) = \mathcal{A}U(t), \ t \ge 0, \\ U(0) = U^0, \end{cases} \tag{4.35}$$

where we have denoted by

$$U(t) = \begin{pmatrix} v(t) \\ u(t) \end{pmatrix} \quad \text{and} \quad U^0 = \begin{pmatrix} v^0 \\ u^0 \end{pmatrix}.$$

On the one hand, since the operator $\mathcal{A}$ is m-dissipative (see [3, section 6] for instance), then according to Theorem 4.2.1, system (4.35) is well-posed as given by Definition 4.2.5. On the other hand, according to Sect. 4.2, system (4.35) is equivalent to the following integral equation:

$$U(t) = U^0 + g_\alpha * \mathcal{A}U(t). \tag{4.36}$$

Equation (4.36) can also be written as follows:

$$\begin{pmatrix} v(t) \\ u(t) \end{pmatrix} = \begin{pmatrix} v^0 \\ u^0 \end{pmatrix} + g_\alpha * \left[\begin{pmatrix} 0 & I \\ -A & -BB^* \end{pmatrix} \begin{pmatrix} v(t) \\ u(t) \end{pmatrix} \right].$$

Equivalently, we have

$$u(t) = u^0 - g_{2\alpha} * Au(t) - g_\alpha * Av^0 - g_\alpha * BB^*u(t), \tag{4.37}$$

where we have used the semigroup property (4.4). By taking $v^0 = 0$ and $u^0 \in X_{\frac{1}{2}}$ such that $BB^*u^0 \in X$ in (4.37), we obtain

$$u(t) = u^0 - g_{2\alpha} * Au(t) - g_\alpha * BB^*u(t). \tag{4.38}$$

Since in this case $U(t)$ is given by the couple $\begin{pmatrix} g_\alpha * u(t) \\ u(t) \end{pmatrix}$, where $u(t)$ is the solution of the system (4.38) and the problem (4.35) is well-posed so by Definition 4.2.5, $U \in C(\mathbb{R}, \mathcal{D}(\mathcal{A}))$ and $g_{1-\alpha} * U \in C(\mathbb{R}, \mathcal{D}(\mathcal{A}))$, this implies that $u \in C(\mathbb{R}_+, X_{\frac{1}{2}})$, $g_\alpha * u \in C(\mathbb{R}_+, X_{\frac{1}{2}})$, and $g_{1-\alpha} * (u - u^0) \in C^1(\mathbb{R}_+, X)$. Moreover, by applying the operator $\mathbf{D}_t^\alpha$ on both sides of (4.38), we find that (4.38) is equivalent to

$$\mathbf{D}_t^\alpha u(t) + g_\alpha * Au(t) + BB^*u(t) = 0, \qquad \forall t > 0,$$

where we have used here the semigroup property (4.4) and (4.9). Now we have proved that system (4.35) with $U^0 = \begin{pmatrix} 0 \\ u^0 \end{pmatrix}$ with $u^0 \in X_{\frac{1}{2}}$ such that $BB^*u^0 \in X$ is equivalent to Eq. (4.38). Since (4.35) is well-posed in the sense of Definition 4.2.5, then problem (4.31) is also well-posed in the sense of Definition 4.4.2 and the regularities of the solution $u(t)$ of (4.31) given by the theorem hold.

Now, if assumptions (4.32) hold true, then according to Theorem 4.3.1 the solution U of (4.35) satisfies the following estimation:

$$\|U(t)\|_{\mathcal{H}} \leq \frac{C}{1+t^{\alpha}} \|U^0\|_{\mathcal{H}}, \quad \forall U^0 \in \mathcal{H}, \ \forall t \geq 0,$$

for some constant $C > 0$, and then we obtain

$$\|u(t)\|_X + \|g_{\alpha} * u(t)\|_{X_{\frac{1}{2}}} \leq \frac{C}{1+t^{\alpha}} \|u^0\|_X, \quad \forall u^0 \in X_{\frac{1}{2}}, \ \forall t \geq 0.$$

This implies in particular the estimates (4.33) and (4.34) and completes the proof.

$\square$

Example 4.4.1 We consider the following integro-differential equation:

$$\begin{cases} \mathbf{D}_t^{\alpha} u(t, x) - g_{\alpha} * \Delta u(t, x) + a(x)u(t, x) = 0, \ t > 0, x \in \Omega, \\ u = 0, \ (0, +\infty) \times \partial\Omega, \\ u(0, \cdot) = u^0 \in H_0^1(\Omega), \end{cases} \tag{4.39}$$

where Ω is a smooth bounded open domain of $\mathbb{R}^n$ and $a \in L^{\infty}(\Omega)$ is non-identically zero and positive function.

By a direct application of Theorem 4.4.1, we obtain a polynomial stability result for (4.39) under a geometric condition G.C.C. (see [37] for more details) on the support of a.

Bibliography

1. Z. Achouri, N.E. Amroun, A. Benaissa, The Euler-Bernoulli beam equation with boundary dissipation of fractional derivative type. Math. Methods Appl. Sci. **40**, 3837–3854 (2017)
2. S. Aizicovici, H. Petzeltová, Asymptotic behavior of solutions of a conserved phase-field system with memory. J. Integral Equations Appl. **15**, 217–240 (2003)
3. K. Ammari, F. Hassine, L. Robbiano, Fractional-feedback stabilization for a class of evolution systems. J. Differ. Equ. **268**, 5751–5791 (2020)
4. K. Ammari, F. Hassine, L. Robbiano, Stabilization for the wave equation with singular Kelvin-Voigt damping. Arch. Ration. Mech. Anal. **236**, 577–601 (2020)
5. K. Ammari, M. Tucsnak, Stabilization of second order evolution equations by a class of unbounded feedbacks. ESAIM Control Optim. Calc. Var. **6**, 361–386 (2001)
6. K. Ammari, F. Hassine, L. Robbiano, Stabilization for vibrating plate equation with singular structural damping. arXiv:1905.13089
7. K. Ammari, S. Nicaise, *Stabilization of elastic systems by collocated feedback* (Springer, Cham, 2015), p. 2124
8. K. Ammari, A. Henrot, M. Tucsnak, Asymptotic behaviour of the solutions and optimal location of the actuator for the pointwise stabilization of a string. Asymptot. Anal. **28**, 215–240 (2001)
9. K. Ammari, A. Henrot, M. Tucsnak, Optimal location of the actuator for the pointwise stabilization of a string. C. R. Acad. Sci. Paris Sér. I Math. **330**, 275–280 (2000)
10. K. Ammari, Z. Liu, M. Tucsnak, Decay rates for a beam with pointwise force and moment feedback. Math. Control Signals Systems **15**, 229–255 (2002)
11. K. Ammari, M. Tucsnak, Stabilization of Bernoulli-Euler beams by means of a pointwise feedback force. SIAM J. Control Optim. **39**, 1160–1181 (2000)
12. W. Arendt, C.J. Batty, Tauberian theorems and stability of one-parameter semigroups. Trans. Am. Math. Soc. **306**(2), 837–852 (1988)
13. E. Bajlekova, *Fractional evolution equations in Banach spaces*, PhD Thesis (Eindhoven University of Technology, Eindhoven, 2001)
14. D. Ben-Avraham, S. Havlin, *Diffusion and reactions in fractals and disordered systems* (Cambridge University, Cambridge, 2000)
15. J.P. Bouchaud, A. Georges, Anomalous diffusion in disordered media: statistical mechanisms, models and physical applications. Phys. Rep. **195**(4–5), 127–293 (1990)
16. C. Bardos, G. Lebeau, J. Rauch, Sharp sufficient conditions for the observation, control, and stabilization of waves from the boundary. SIAM J. Control Optim. **30**, 1024–1065 (1992)
17. C.J.K. Batty, T. Duyckaerts, Non-uniform stability for bounded semi-groups on Banach spaces. J. Evol. Equ., 765–780 (2008)

© The Author(s), under exclusive license to Springer Nature Switzerland AG 2022
K. Ammari et al., *Stabilization for Some Fractional-Evolution Systems*,
SpringerBriefs in Mathematics, https://doi.org/10.1007/978-3-031-17343-1

18. A. Borichev, Y. Tomilov, Optimal polynomial decay of function and operator semigroups. Math. Ann. **347**(2), 455–478 (2010)

19. H. Brezis, *Functional analysis, Sobolev spaces and partial differential equations* (Springer, New York, 2011)

20. N. Burq, Décroissance de l'énergie locale de l'équation des ondes pour le problème extérieur et absence de résonance au voisinage du réel. Acta Math. **180**(1), 1–29 (1998)

21. L. Caffarelli, C. Chan, A. Vasseur, Regularity theory for parabolic nonlinear integral operators. J. Am. Math. Soc. **24**(3), 849–869 (2011)

22. L. Caffarelli, A. Vasseur, Drift diffusion equations with fractional diffusion and the quasi-geostrophic equation. Ann. Math. **171**(3), 1903–1930 (2010)

23. L. Caffarelli, S. Salsa, L. Silvestre, Regularity estimates for the solution and the free boundary of the obstacle problem for the fractional Laplacian. Invent Math. **171**(2), 425–461 (2008)

24. L. Caffarelli, L. Silvestre, An extension problem related to the fractional Laplacian. Commun. Partial Differ. Equ. **32**(7–9), 1245–1260 (2007)

25. M. Caputo, M. Fabrizio, A new definition of fractional derivative without singular kernel. Progr. Fract. Differ. Appl. **1**, 73–85 (2015)

26. J.W.S. Cassals, *An introduction to Diophantine Approximation* (Cambridge University Press, Cambridge, 1966)

27. P. Constantin, A. Kiselev, L. Ryzhik, A. Zlatoš, Diffusion and mixing in fluid flow. Ann Math. **168**(2), 643–674 (2008)

28. E. Cuesta, Asymptotic behaviour of the solutions of fractional integro-differential equations and some time discretizations, in *Discrete and Continuous Dynamical Systems* (2007), pp. 277–285

29. A. Erdélyi, W. Magnus, F. Oberhettinger, F.G. Tricomi, *Higher Transcendental Functions*, vol. III (McGraw-Hill, New York, 1955)

30. R. Gorenflo, Y. Luchko, F. Mainardi, Analytical properties and applications of the Wright function. Fractional Calculus and Applied Analysis **2**(4), 383–414 (1999)

31. S. Das, *Functional fractional calculus for system identification and control* (Springer, Berlin, 2011)

32. J.F. Gómez-Aguilar, D. Baleanu, Solutions of the telegraph equations using a fractional calculus approach. Proc. Romanian Acad. Ser. A **15**, 27–34 (2014)

33. F. Hassine, Remark on the pointwise stabilization of an elastic string equation. Z. Angew. Math. Mech. **96**(4), 519–528 (2016)

34. F. Huang, Characteristic conditions for exponential stability of linear dynamical systems in Hilbert space. Ann. Differential Equations **1**, 43–56 (1985)

35. A. Kiselev, F. Nazarov, A. Volberg, Global well-posedness for the critical 2D dissipative quasigeostrophic equation. Invent. Math. **167**, 445–453 (2007)

36. S. Lang, *Introduction to diophantine approximations* (Addison Wesley, New York, 1966)

37. G. Lebeau, Equation des ondes amorties, in *Algebraic and geometric methods in mathematical physics* (Springer, Dordrecht, 1996), pp. 73–109

38. J.A. Machado, I.S. Jesus, R. Barbosa, M. Silva, C. Rei, Application of fractional calculus in engineering. Dynamics, Games and Science I **1**, 619–629 (2011)

39. J.A.T. Machado, A.M. Lopes, Analysis of natural and artificial phenomena using signal processing and fractional calculus. Fract. Calc. Appl. Anal. **18**, 459–478 (2015)

40. R.L. Magin, *Fractional calculus in bioengineering* (Begell House, Redding, 2006)

41. F. Mainardi, Fractional relaxation-oscillation and fractional diffusion-wave phenomena. Chaos, Solitons Fractals **7**(9), 1461–1477 (1996)

42. D. Matignon, C. Prieur, Asymptotic stability of Webster-Lokshin equation. Math. Control Relat. Fields. **4**, 481–500 (2014)

43. D. Matignon, C. Prieur, Asymptotic stability of linear conservative systems when coupled with diffusive systems. ESAIM Control Optim. Calc. Var. **11**, 487–507 (2005)

44. B. Mbodje, Wave energy decay under fractional derivative controls. IMA J. Math. Control. Inf. **23**, 237–257 (2006)
45. B. Mbodje, G. Montseny, Boundary fractional derivative control of the wave equation. IEEE Trans. Autom. Control **40**, 368–382 (1995)
46. R. Metzler, J. Klafter, The restaurant at the end of the random walk: recent developments in the description of anomalous transport by fractional dynamics. J. Phys. A **37**(31), 161–208 (2004)
47. A. Pazy, *Semigroups of linear operators and applications to partial differential equations* (Springer, New York, 1983)
48. I. Podlubny, *Fractional Differential Equations* (Academic Press, San Diego, 1999)
49. J. Prüss, *Evolutionary Integral Equations and Applications* (Birkhäuser, Basel, 1993)
50. J. Prüss, On the spectrum of C_0-semigroups. Trans. Am. Math. Soc. **284**, 847–857 (1984)
51. J. Prüss, V. Vergara, R. Zacher, Well-posedness and long-time behaviour for the non-isothermal Cahn-Hilliard equation with memory. Discrete Contin. Dyn. Syst. **26**, 625–647 (2010)
52. L. Qi, E. Zuazua, On the lack of controllability of fractional in time ODE and PDE. Math. Control Signals Syst. **28**(2), 10 (2016)
53. S.G Samko, A.A. Kilbas, O.I. Marichev, *Fractional integrals and derivatives, Theory and applications* (Gordon and Breach Science Publishers, Yverdon, 1993)
54. R. Stahn, Optimal decay rate for the wave equation on a square with constant damping on a strip. Z. Angew. Math. Phys. **68**(2), Art. 36, 10 pp. (2017)
55. V.E. Tarasov, *Fractional Dynamics: Applications of Fractional Calculus to Dynamics of Particles, Fields and Media* (Springer, Berlin, 2011)
56. L. Tébou, A constructive method for the stabilization of the wave equation with localized Kelvin-Voigt damping. C. R. Acad. Sci. Paris Ser. I **350**, 603–608 (2012)
57. L. Tébou, Stabilisation of some elastic systems with localized Kelvin-Voigt damping. Discrete Contin. Dynam. Systems **36**, 7117–7136 (2016)
58. M. Tucsnak, On the pointwise stabilization of a string, in *Control and Estimation of Distributed Parameter Systems*, vol. 126. International Series of Numerical Mathematics, vol. 145, eds. by W. Desch, F Kappel, K. Kunisch (Birkhäuser, Basel, 1996), pp. 287–295
59. M. Tucsnak, G. Weiss, *Stabilization of elastic systems by collocated feedback* (Birkhäuser, Basel, 2009)
60. D. Valério, J.A.T. Machado, V. Kiryakova, Some pioneers of the applications of fractional calculus. Fract. Calc. Appl. Anal. **17**, 552–578 (2014)
61. M. Walter, *Dynamical Systems and Evolution Equations, Theory and Applications* (Plenum Press, New York, 1980)
62. E.M. Wright, The generalized Bessel function of order greater than one. Q. J. Math. (Oxford Ser.) **11**, 36–48 (1940)

Printed in the United States
by Baker & Taylor Publisher Services